What are people saying about...

Mind, Quantum, and Free Will

Mind, Quantum, and Free Will is engagingly written, with an impressive breadth and depth of scholarship. I found it quite provocative in multiple respects, and believe that most readers with interests in these topics would be enriched by exploring it.
Andrew Westcombe, PhD in philosophy; specialist in metaphysics and the mind-body problem

Previous publications

Ells, P. (2011). *Panpsychism: The philosophy of the Sensuous Cosmos* (Winchester: John Hunt Publishing)
ISBN 978-1-84694-505-2

Ells, P. (2018). "Alternatives to Physicalism", in Castro, J., Fowler, B. & Gomes, L. (Eds.), *Time, Science and the Critique of Technological Reason: Essays in Honour of Hermínio Martins* (Switzerland: Palgrave Macmillan)
ISBN 978-3-319-71518-6

Mind, Quantum, and Free Will

The Birth of Physics in the Sensuous Cosmos

Mind, Quantum, and Free Will

The Birth of Physics in the Sensuous Cosmos

Peter Ells

Winchester, UK
Washington, USA

JOHN HUNT PUBLISHING

First published by iff Books, 2022
iff Books is an imprint of John Hunt Publishing Ltd., No. 3 East Street, Alresford,
Hampshire SO24 9EE, UK
office@jhpbooks.com
www.johnhuntpublishing.com
www.iff-books.com

For distributor details and how to order please visit the 'Ordering' section on our website.

ISBN: 978 1 78535 965 1
978 1 78535 966 8 (ebook)
Library of Congress Control Number: 2022930126

A CIP catalogue record for this book is available from the British Library.

Design: Matthew Greenfield

UK: Printed and bound by CPI Group (UK) Ltd, Croydon, CR0 4YY
Printed in North America by CPI GPS partners

Contents

For my parents
Dorothy and Jack

Many thanks to my wife, Elena
For supporting me in my writing

Why this book?

The *mind-body problem* is the most important, exciting and intractable enigma that humankind has ever faced. It has perplexed thinkers over many centuries, and has come under intense scrutiny in recent decades. How can mind (which possesses features that include thinking, feeling, emotion, perception, and free will – the ability within limits to choose which action one takes) possibly be consistent with the essential character of the universe, which is nowadays understood to be a physical system (comprising forces, particles, waves, space and time, physical laws, and suchlike)? There seems to be no way to explain mental features in physical terms – mentality appears to be a complete novelty, irreducible to physics.

One might think that, as the sciences have progressed over recent decades, the problem would resolve itself. This is not so: if anything – as Part I will show – the rapid advances and great successes of science have made the problem even more acute. Physics gives the completest possible description of goings-on in the world, without any need of mentalistic terms. (There is a word of caution here that will be explained later. The bottom line is that, although the arguably-mentalistic concept of 'measurement' features in quantum mechanics, this is of no immediate help in solving the mind-body problem.) Mind would therefore seem to be superfluous. Nonetheless, some obstinate facts remain – **we do experience the world**: we feel pain when we break our leg; we feel the warmth of the Sun on our faces; we see a tree nearby; we sometimes feel thirsty; and so on. Moreover, **we are agents who make a difference to what goes on in the world**. Your decision to quench your thirst instigates the sequence of events in which a cup of coffee is prepared and drunk – these physical changes taking place in your kitchen are caused by your mind.

The ideas presented in this book are significant advances

towards solving the mind-body problem; and they are consistent with current scientific understanding. Moreover, the theory allows almost all of the claims in the above sketch to be literally true. In contrast, most attempts to solve the mind-body problem assert that many of these seemingly self-evident claims (that we have genuine free will for example) are altogether misguided. This is why I have written the book, and you might be interested in reading it. Some specific advances are listed here:

- *The mind-body problem is not a scientific problem but a philosophical one*

The entire project, of attempting satisfactorily to explain pain – an incontrovertible fact of experience – in terms of particular physical events occurring within the brain, is misguided.

- *A better philosophical basis for posing the question is given*

This basis combines idealism (the claim that everything which exists has, at its root, the character of mind) with realism about the universe and its contents.

- *The solution given here is precisely linked to physics*

This well-defined link is a significant advance on anything that has been achieved previously.

Each point will be outlined in turn.

The mind-body problem is not a scientific problem but a philosophical one

A way of seeing this is to note, as David Chalmers did at a conference in 1996(b), that we do not have a consciousness meter – a piece of apparatus which, when pointed at an object such as a rock or a brain, detects whether or not it is having a conscious experience, for example that of feeling cold. He famously used his hairdryer to represent such a meter. No one has any inkling as to how such a device might be developed – even in principle. In particular, how is one to check, using scientific methods and assumptions alone, whether or not such a device is working correctly? If the ideas presented in this book are correct, consciousness meters are impossible.

In our everyday lives, we use empathy (grounded in our personal history) to decide what another person is experiencing in a given situation; this judgement is based on their behaviour and other physical evidence. For example, we see someone shivering in the snow and realise they are feeling cold. But we do not suppose that their experience is **nothing other than** this overt behaviour. Nor do we have any reason to believe that experience is identical to covert behaviour – the physical goings-on within their brain. The position developed in Part III confirms that the intuitions of this paragraph are factually true (even though most academics have claimed that they are necessarily false).

The particular case made in this book is that the mind-body problem is a metaphysical one. *Metaphysics* is the branch of philosophy which catalogues the fundamental entities that exist in the universe; and which describes their essential character (or nature), together with their properties. There may be additional, secondary things that exist, but if so, these extras are to be fully accounted for in terms of the basic entities.

The success of science, in particular of physics, has led many to adopt the metaphysical position of *physicalism*. This asserts that the fundamental entities of the world are physical (electrons, magnetic fields, and suchlike); and that all of the facts of the world are grounded in truths about physical entities, properties and laws. An example of a non-basic entity is a monkey: according to physicalists, all of its properties can be explained fully in terms of physical stuff, properties and laws, including importantly the particles and fields of which it is composed. This approach, however, leads one to a dead-end when attempting to solve the mind-body problem.

It turns out that the root of the difficulty of the mind-body problem lies in the unwarranted assumption that – in order to be consistent with well-established scientific truths – the only possible metaphysical system is physicalism. This assumption is made without sufficient critical reflection by most scientists, and even by some philosophers. It is not unusual to see the mind-body problem expressed as: "How can we explain all of the characteristics of mind in physical terms?" (This is more-or-less how the problem was phrased in the very first paragraph above.) However, by putting the problem in this manner, a writer is unconsciously assuming the truth of physicalism and, being thus self-constrained, is doomed to fail: No physicalist account of mind has come close to giving a satisfactory account of experiential qualities, such as pain, or the experienced blueness of a clear sky. Moreover, as will become clear, no variant of physicalism is consistent with the undeniable reality of human agency. Numerous variants of physicalism have been proposed, as we shall see in Part I, but all of them fare badly in solving the mind-body problem. However, if we reject physicalism, the problem then becomes tractable.

A better philosophical basis for posing the question is given

The alternative approach taken in this book is based on *idealism*, which is the metaphysical position that takes minds, and things pertaining to minds (such as experiences, percepts, qualities, free will, agency, and so on), to be the fundamental things that exist. Under this approach, the *idealist mind-body problem* becomes that of **deriving completely all of the physical facts of the world in mentalistic terms**. There are many variants of idealism (just as there are with physicalism), but most of them give an inadequate account of the mind-body problem. A broad introductory survey may be found in *Idealism: The History of a Philosophy* (J. Dunham & others, 2011).

The particular variant of idealism that I study in Part III of this book, called *pan-idealism,* is unusual in that it is thoroughly realistic about the entities in our universe (realism about electrons, flowers, drops of water, galaxies, and so on). It is fairly clear that unrealistic idealisms are unlikely candidates for providing solutions to the mind-body problem. It will be shown that the list of entities which exist in the universe according to pan-idealism is essentially identical to the list that would be provided by physicalists.

The prefix "pan" comes from *panpsychism,* the idea that mind or mental properties are present everywhere throughout the cosmos. Usually it is taken to be a kind of physicalism, in which every basic entity ***has* mental properties**. In pan-idealism we make a stronger claim: basic entities are omnipresent, and every one of them ***is* a mind**, albeit typically extremely primitive.

A minority of readers will already be familiar with panpsychism. It is an attractive position and solves many aspects of the mind-body problem. A remaining difficulty is the *combination problem*: even if we accept that the fundamental particles of physics each have a primitive mentality, how could they possibly, when they come together to form a human body, also combine to form a human mind, which has a necessary (albeit imperfect) unity of experience? This problem is widely known, and about half the papers written by panpsychists are devoted to attempting to solve it. Pan-idealism solves the combination problem of panpsychism, as will be seen in chapter 10.

The solution given here is precisely linked to physics

The mind-body problem for pan-idealism is to give a full account of the physics of the world – consistent with current-day science – in terms of this mentalistic foundation. There are several principles that help achieve this, and all will be detailed in Part III:

- (Already mentioned) The catalogues of entities that exist, according to our current understanding of physics, and to pan-idealism, are identical

There is a slight complication here, because physics does not provide such a catalogue as unambiguously as one might hope. Explicit definitions of the two catalogues will be given in chapters 8 and 10.

- Our sole knowledge of the physical world comes to us through our experiences and our theories – which are both mentalistic in character

Pan-idealism leverages this circumstance, plus the fact that primitive minds are ubiquitous throughout the cosmos, to **identify** each physical fact with a mental fact – one that is obtained by combining, as consistently as possible, the percepts of these primitive minds. This combination uses much the same principles that scientists employ when working together on an experiment: they reconcile their individual experiences to give a single, coherent account of what is going on physically. There are differences: primitive minds, despite their percepts, have no intellectual powers; on the other hand, they are innumerable and omnipresent. The identification process will be explained in chapter 8.

- Pan-idealism gives a realist account of physical entities perceived in experience

Surprisingly, pan-idealism gives a more explicit account of what it is to be real than does physicalism; see chapters 7-8. The bare bones of the argument are these: According to physicalists, a drop of water is real if it boils like a droplet, freezes like a droplet, flows down a windowpane like a droplet, etc. Moreover, there is the further requirement that other observers examining the phenomena witness similar results. Notice that all these criteria for the reality of the droplet are empirical – they are based solely on what is going on in the minds of these observers.

Pan-idealism accepts all of the above, but it makes an additional requirement: In order for the water droplet to be real, its constituent particles (and perhaps the drop itself) must be centres of experience. This is a definitive distinction between a real water drop and a collective hallucination. The objection might be raised that this distinction is untestable. But, given that we are finite creatures, it is immodest to suppose that there are no truths about the universe that will be forever beyond our grasp.

- Pan-idealism is consistent with free agency

In contrast to all forms of physicalism (including panpsychism), pan-idealism has no problems in explaining human free will, and also the simpler forms of agency in more primitive entities. This is because agency is the **only** form of causation according to pan-idealism. Agency, in some form, cannot intelligibly be denied: without it, consciousness is at best an ineffectual appendage to the universe. If I didn't possess agency, you would have no reason to believe that the words in this book in any way reflect my thoughts: denying agency amounts to denying consciousness.

- Pan-idealism is closely tied to current, fundamental physics

Pan-idealism can be closely tied to any up-to-date, observer-free interpretation of quantum mechanics. The mind-body problem is a philosophical one and not a physical one. For this reason, we do not want mind to **appear within** the physics of the world – instead the goal is to **derive** physics **from** mind. A particular interpretation of quantum mechanics, called GRW, is suitable, and is examined in chapter 10, but pan-idealism is readily adapted to other observer-free interpretations, including Roger Penrose's OR theory (1989, 1995).

Points of clarification

In this book I focus on commonplace waking states of consciousness, as in the example of making a cup of coffee. I briefly touch upon altered states of consciousness, such as the changes one experiences after drinking a glass of wine. There is no mention of alleged paranormal phenomena such as extrasensory perception (about which I am sceptical). My attitude is that, before one attempts to investigate such matters, a better understanding of the everyday mind-body relationship is required.

Some readers might wish to reject pan-idealism as absurd, citing *Occam's razor*, which states that "Entities should not be multiplied without necessity." My reply is that, though it possesses some seemingly absurd features, pan-idealism is an entirely rational system. Moreover, these features are a necessity if we are to solve the mind-body problem.

Physicists may find the novel two-stage dynamical process within GRW, presented in chapter 10, to be of value. This applies even to those who entirely reject my approach to the mind-body problem.

Some experts offer the opinion that "Quantum mechanics is a mystery; consciousness is a mystery. Those who investigate mind in terms of quantum mechanics are just combining two mysteries into one." Such people will reject my book just on looking at the title. I have two responses. First, setting aside the Copenhagen interpretation (chapter 5), quantum mechanics is not mysterious. Second, what is presented here is a specific theory, worked out in depth.

Outline

The book is structured as follows:

Part I *presents the mind-body problem.* Chapter 1 explains this, and describes the failures to date to give an adequate physicalist account of mind. Chapter 2 introduces some relevant philosophical ideas. Towards the end, a Table gives my imprecise evaluations of four metaphysical systems in tackling aspects of this problem.

Part II *summarises the fundamental theories of physics as these are currently known.* Chapter 3 is devoted to classical physics, including relativity, and chapters 4-6 to quantum theory and its interpretations.

Quantum theory is presented in a standard way here, at least in terms of its physics. Of interest is that I've also taken care to describe a cultural disagreement that broke out among scientists in the twentieth century. The majority, led by Niels Bohr of the Copenhagen School, understood the theory to mean that scientists should no longer attempt to explain how the world worked; instead they should henceforward restrict themselves to predicting the outcomes of experiments. A minority, most notably Einstein and Schrödinger, rejected this watering down of the ambitions of science.

The Copenhagen School reigned supreme, and Einstein was belittled, this notwithstanding his eminence, and his eleven major contributions to quantum theory (one of which earned him the Nobel Prize). In this century, Copenhagen is still dominant, though less so than before. Einstein's reservations about the adequacy of quantum theory in explaining the world (he agreed that it was superbly accurate in predicting experimental outcomes) was creatively useful. His qualms eventually led to a spectacular new scientific finding about the universe – that it is *entangled*. This novel feature was not demonstrated until many years after the Copenhagen School had insisted that the theory had already been perfected. Part II defends Einstein and Schrödinger's approach, showing it to be far more fruitful than the ill-defined complacency of the Copenhagenites.

Part III *is my solution to the mind-body problem*. Chapter 7 argues that the concept of concrete (versus abstract) reality does not belong within the ambit of science; instead, this distinction is factual, but belongs within the domain of philosophy. Chapter 8 introduces pan-idealism. Chapter 9 discusses free will and its relation to indeterminacy. Here, an important critique by Schrödinger is answered. Chapter 10 describes centres of experience in terms of a specific quantum mechanical model (GRW). (Penrose's OR theory is also discussed here.) Pan-idealism is thus closely tied to modern physics. Chapter 11 sums up.

Part I

The mind-body problem

Chapter 1

The problem

This introduces the mind-body problem, and explains why philosopher David Chalmers calls the intractability of its core issues "The Hard Problem". Nearly all scholars attempt to tackle the problem within a particular general framework. This goes by the name *physicalism* because it asserts that physics is fundamental: everything that is not physics – biology for example, or even economics – can be boiled down to physics. Physicalists have developed various distinct accounts of how mind arises in the physical world; and a representative selection will be discussed and critiqued here. Most studies have focussed on perception – the predominantly passive experience of the world about us. Much less attention has been given to agency and free will – the active role of mind in the world.

The conclusion reached by the end of the chapter is that every physicalist approach to the mind-body problem has failed badly – at least up to the present day. Moreover, there is no prospect that any physicalist solution will be successful in the foreseeable future. The situation thus appears bleak, but we have yet to consider the possibility of working in an alternative *non-physicalist* framework (i.e., one in which all physical facts are to be taken as secondary, rather than fundamental). In developing such an approach, we must of course retain respect for science's ability to describe the universe correctly. A non-physicalist framework might remove the intractable "Hardness" of the mind-body problem, even if it doesn't solve it completely. This hope leads us on to succeeding chapters.

The mind-body problem

Figure 1.1 shows a simplistic picture of the mind-body problem as it initially presents itself to us. There is a story that goes along with it: **C**) Your plant sits in the greenhouse, parched and drooping. **A**) You enter the greenhouse and see it. **B**) You have the idea that if you water the plant, it will revive. **D**) You fill a glass with from a nearby tap and pour it over the plant, which perks up.

At first sight there seem to be two 'worlds', the physical world and the mental:

The physical world (**C** and **D**) has an excellent description in terms of physical entities, properties and laws. We have a scientific understanding of why the plant revives when water falls on it. We can also calculate how fast the drops of water will fall. **D** occurs later than **C**, and dynamical physical laws provide a causal explanation as to why we arrive at **D** from **C**.

Your mental world (**A** and **B**) is rather different. **A** is your percept that your houseplant is drooping. Percepts are replete with *qualities*: these are the colours, **as these are experienced**, of the plant and pot – the greenness of the stem, for instance; the scent of the flower or your feeling of warmth give further examples. Each such experiential quality has the formal Latin name *quale* (pronounced kwa-lay; plural *qualia*). Qualia are essentially mental phenomena: by definition there can be no qualia in the absence of a mind. Take for example the perfume of the flower. A partial explanation for this is that the flower emits molecules with a specific structure, and these enter your nose. This eventually leads to some goings-on in your brain. These in turn give rise to your experience, including the quale that is your experience of the perfume of the flower. But there is no explanation of qualia in terms of physics. There is a seemingly complete and acceptable scientific description of such molecules and their effects on brain dynamics – even though physics textbooks do not mention, let alone explain, qualia.

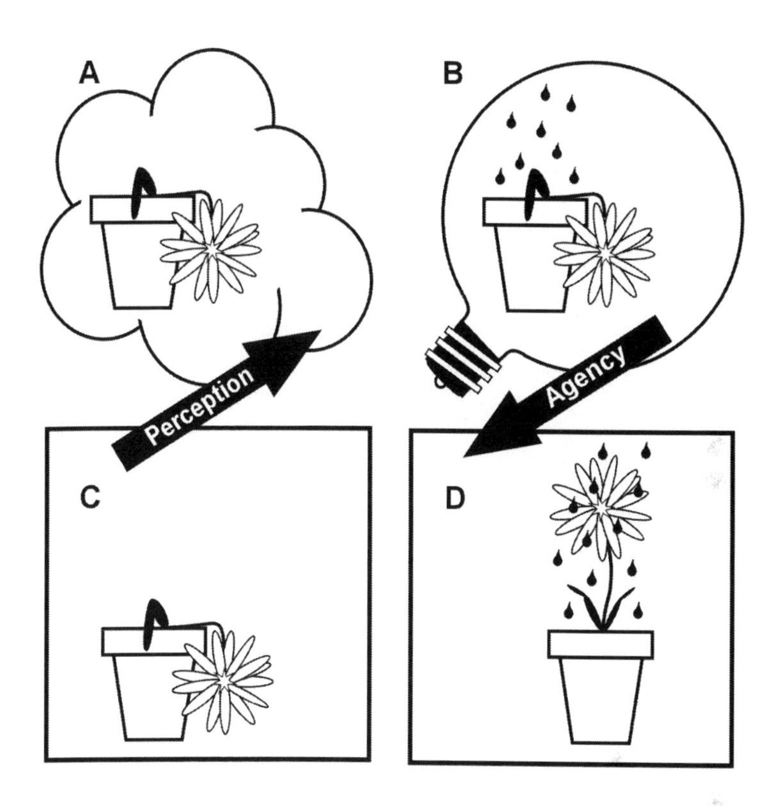

Figure 1.1: A simplistic picture of the mind-body problem

B represents your idea that if you water the plant, then it will revive. Having had any idea, you have a free choice as to whether or not to carry it into effect. In this case you decide to water the plant. You possess *agency*, which means that, by a mental act, you can move your hands so as to fill a glass with water from a nearby tap and pour it over the plant.

The mind-body problem thus arises because there are two pictures we can sketch of a human person, both of which we have excellent reasons to believe in, but which have been extraordinarily difficult to reconcile – even after being studied by the finest intellects over centuries.

First, regarded in physical terms, a human person is a *biological organism* – existing as a minuscule fragment of the physical universe. There can be little doubt that every cell and organ, brain included, of this complex body obeys exactly the same spatiotemporal laws as any other physical system. Let us compare various objects in our world. The differences between an adult human, a foetus, an insect, a plant, a bacterium, a virus, and inorganic material such as a drop of water or a pebble – *measured solely in terms of structure and behaviour* – are merely differences in complexity.

Second, a human person exists as a *mind* that experiences the world. Human percepts reveal the world to us. Indeed, our entire knowledge of the universe and its contents comes to us through our experiences – and in no other manner. Some things, such as items of furniture, are given to us more-or-less immediately in our experiences. Other things, such as atoms, are theoretical inferences, derived from humankind's multitude of experiences of taking readings from scientific instruments.

There are three types of links between these pictures, all of which are perplexing:

First there is **perception**. As drawn in the simplistic picture, the perceived flower is identical to the flower itself. This naive picture cannot be true, but we have reason to believe that our percept is an adequate representation of those facts about the flower that are most useful to us for our survival: is the flower recognisable as one that is poisonous or good to eat, for example. Qualia are the most puzzling aspect of our percepts, as they are not present in the physical world. They do, however, provide us with useful information: from its golden colour, a farmer can know that the wheat crop is ripe for harvesting.

As minds, moreover, we seem to be **agents** in the world: I decide to write the word 'bedazzle' and the necessary sequence of movements were produced by my fingers: physical events are produced by mental events. Going further, do we possess **free will**: are my decisions freely chosen by me? In the case of the arbitrary word, perhaps my choice was not free. But, in the case of an author writing a book, whether or not to begin, the topic decided upon, and the words to be used, are all, prima facie, free, conscious, individual choices. The onus is on those who deny that we possess free will to give a compelling argument here. The cases in which the choices strongly define who one is as a person (as in writing a book, who to marry, what career path to follow) are those for which it is most difficult to argue against authentic free will. For trivial ('plant watering') or arbitrary ('bedazzle') examples of choices, the argument against free will is more readily made. Psychological experiments typically focus on these latter types of choices.

Linking the above two viewpoints are the strong **correlations** between observed physical brain states and reported experiences. Included in this are drugs that affect states of wakefulness or tiredness; elevation or depression of mood; and so on. Damage to the brain, whether caused by trauma, illness, or old age, almost invariably results in an impaired intellect. The strength and dose of the drug in the first example, and the extent of damage in the second, are in general proportionate to changes in the person's experiences. Conversely, people reporting a changed mood such as elation or depression, wakefulness, or tiredness, if examined by (say) a CT scan, quite often manifest differences in their physical brain activity. Two equivalent ways of describing these links is to say that there are *neural correlates of consciousness* (NCCs), or that there is *psychophysical parallelism*. In this book, these expressions are just descriptions of the observed facts, and are neutral as to any explanation that may later be given.

The mind-body problem thus has several facets:

1. *At first sight there are two, contradictory worlds: the physical and the mental*

  1.1. *Which world are we living in?*

  1.2. *What is the relationship between a person's mental state and their brain state?*

2. *How can minds perceive the physical world?*
   - 2.1. *How do structural physical facts (a book is in front of you) become structural mental facts (your visual percept is of a book)?*
   - 2.2. *How can we explain qualia?*

3. *How can minds, as agents, cause changes in the physical world?*
   - 3.1. *How are mental and physical causation related?*
   - 3.2. *Can we make free choices? If so, how?*

4. *How did consciousness evolve in a universe that was initially lifeless for billions of years?*

Reconciling the scientific and experiential viewpoints is extraordinarily difficult; but first it is necessary to describe each world and the links between them in more detail.

Experience

The first viewpoint is that of the experiences and activities of our everyday lives. At the moment I am sitting at my desk. Looking around I can see various items nearby. There is a pale blue hole-punch with a plastic base that feels soft to the touch. Adjusting my anglepoise lamp I feel the resistance of its springs; and flinch from the burning sensation when I touch its shade. I can see the yellow glow emitted by its bulb. In my mouth I can still experience the taste of the tea that I drank five minutes ago. I am free to turn my attention where I desire, but I am also free to leave off describing my surroundings for another activity such as going for a walk.

Writing is an activity that involves perception, memory, imagination and thought. My topic is strongly influenced by my previous plans for this section of the book. The particular examples I gave and the words I used to express them were chosen as I wrote. Fruity tropes @svsv juxtaposition. But I am also free to write nonsense at a whim.

Like yours, my experiences are rich in qualia. Indeed, I could not recognise my hole-punch for what it is in the absence of contrasting visual qualia within my percept, which inform me of its structure. Without them, I would experience the punch as no more than a uniform patch of colour. Qualia are not merely colours, sounds, tastes, and so on; they are *these qualities as they are experienced by an individual*; examples of qualia include experiences of the dryish bitter taste of tea and of the jangling of a bell. Our experiences are so closely tied to physical phenomena that we often use the same word for both. For example, light of a certain frequency typically corresponds to a particular visual experience, and we describe both the light and the experience as being 'blue'. 'Chanel No. 5' is the name given to a specific liquid in a bottle, but it is also given to the name of our experience on sniffing it. A quale is always the experience.

Our percepts of the world have a certain holistic character. When I look at my desk lamp I can almost at once see it for what it is, and understand its function. Most humans can recognise individual faces almost instantly. But some people with a certain type of brain damage, although they can recognise every feature of another's face, have to use reason in order to put these features together and so identify the other person. This rational effort is slow and unreliable. However, the holistic character of even an uninjured person's percept is much more limited than we might at first think. Psychological experiments have shown that we do not take in the entirety of the information falling on our retinas in one go. Instead, what we see is assembled from information taken from an extremely narrow region at the centre of the retina (the fovea). The major part of the retina surrounding the fovea is used to pick out features of the observed scene, such as contrast and movement, so that the eye rapidly and repeatedly jumps (saccades) so as to bring points of interest into the fovea. What we perceive has been constructed from these fragments, and is also highly conditioned by our expectations (Gregory,

1998, pp. 44-50). For example, we clearly recognise a cat walking in the road, but as we continue to look it transforms itself into a black plastic bag blown by the wind.

By a good part of a second, our percepts lag behind events going on in our surroundings. We are usually completely unaware of this fact but, if we accidentally nudge a cup off of a table, and then move to grab it, we find that our hand is inexplicably frozen in space, before it moves uselessly just as the cup smashes on the floor.

From the above discussion it is clear that I do not claim that our naive intuitions about our experiences are unquestionable. A final example is that there is good evidence to suggest that humans might each possess multiple streams of consciousness. Perhaps the strongest is provided by patients who have undergone 'split brain' surgery (formally called partial or complete callosotomy), which separates their cerebral hemispheres, in order to cure their epilepsy. Early in recovery, the two halves of the patient's brain (each half controls the opposite side of the body) can be observed to be in conflict with one another. Perhaps one hand is fastening buttons on a jumper that the patient is wearing, while the other hand is undoing them. Daniel Dennett (1991, chapter 9) has a well-developed "Fame in the brain" theory involving multiple streams of consciousness. The stream that is dominant in determining behaviour at any given moment depends on the task currently being undertaken, but this may be subject to interruption by external stimuli. I may be hastening to the shops to buy food, for example, when I encounter a friend and we both agree to first go to a café and have a chat. Dennett's theory may well be right, but I reject his contention that consciousness is nothing other than information processing in the brain.

Our everyday self-understanding is that we are agents, in the sense that we have the power (within limits) to act on our choices, thus altering the course of events in the world.

Examples can be trivial. When I decide to make a cup of tea, several events occur in the kitchen at my instigation, and among the end results is that there is one less teabag in the jar. But, as already discussed, choices can be far more significant (for example career path); and in these situations, the intuition that we are agents with free will is exceptionally persuasive, and requires strong counter-arguments.

Each of us is a *subject* of experience: I have my experiences of my environs, and you have yours. When we are near one another, our several experiences are similar enough that we can communicate and work together constructively.

Characterising consciousness and experience

In this book, *consciousness* will be understood extremely broadly, in terms of Thomas Nagel's characterisation that an entity is conscious if and only if **"there is something that it is like to be"** that entity (1979, chapter 12 & p. 168). This is not an attempt at a definition. You are familiar with being conscious in your everyday waking life (and in your dreams). If this were not the case, then the emphasized phrase would have no meaning to you – even if you were to look up every word in a dictionary, and study it long and hard. But you **are** familiar with what it is to be conscious, and you have been since you were a child.

Consider pain. We are all familiar with this feeling. Without doubt, when we are in pain, there is something it is like to be us: we exist painfully, and a severe pain can blot out all other experiences. Pain has its own characteristic nature that cannot be doubted. Moreover, it cannot be described adequately in terms of anything else that is not an experience (mild pain is analogous to strong heat or cold, but these are also experiences).

Some writers make a distinction between consciousness and experience. To be conscious, they say, is not only to have experiences, but to be able to reflect upon, think about, and draw conclusions from those experiences. So the claim might perhaps be made that a creature, for example a worm, has bare experiences but is not conscious: It might exist painfully and be in a state of pain, but it would not understand or reflect upon this experience.

The distinction is indeed useful, but my opinion is that, once you have solved the mind-body problem for a non-reflective creature such as a worm, then you have gone most of the way to explaining reflective consciousness and agency in more complex creatures, such as ourselves – or octopuses. For example, an account can be given about a baby learning to recognise its own body, in terms of learning to distinguish which portion of the physical world is under the control of its will. Textbooks on child development would remain unaltered, but perhaps with some of the mentalistic terms being taken more literally.

Often, I will treat the phrases *being conscious* and *having experiences* as synonymous. This is just my convention: it isn't a substantive argument about reality. If I do need to make a distinction, I will contrast *bare experience* with *reflective consciousness* as in the example above. Likewise, I will write of *bare agency* and *reflective agency*. The respective terms always go together: Bare experience and bare agency are found in a worm. Similarly, reflective experience and reflective agency always go together, say in a chimp.

Science as a human endeavour

We now turn to the second viewpoint appearing in Figure 1.1, that of the physical world (**C** and **D**). This section outlines the changing character of scientific theories, and how scientists came to develop them.

From the viewpoint of science, a human person is a biological organism that obeys the same physical laws as any other system or entity within the universe. At least in principle, the behaviour of any physical system has its fullest-possible explanation in terms of the laws which govern the universe. This assertion requires some fleshing out and qualification.

In 1687 Newton published his *Principia*. This work acted as the epitome for the way the physical world was to be understood – at least until the twentieth century. In the theory, space was defined as a three-dimensional geometric structure, and time was conceived of as flowing uniformly throughout the entire universe. Newton proposed deterministic mathematical laws that apply to every entity that exists. His theory could explain the fine detail of the orbits of the planets, including their elliptical paths, their velocities, and even their mutual perturbations (at least insofar as these could be measured in his day). Exactly the same laws could explain the trajectories of balls, and the peculiar processional behaviour of gyroscopes. The laws are *universal*, being true for all places and times in the universe.

(For religious reasons Newton and many of his contemporaries believed that humans were to a certain extent exempt from his laws. Nowadays, nearly all scientists, including many of those with a religious faith, maintain that scientific laws should hold universally and without exemption. This majority view, which I accept, will be assumed throughout this book.)

In 1905 Einstein published his theory of special relativity. By 1916 he had extended this to general relativity, which now included an account of gravitation. Einstein's theory is almost identical to Newton's in situations where the speeds of objects are small compared to the speed of light, and where objects are not too heavy. In situations where these constraints do not apply, Einstein's model is much more accurate than Newton's, which breaks down completely. Einstein's theory has broadly the same character as Newton's. It has a geometric model of space and time, which must now be considered to be melded together in a curved four-dimensional unity, called spacetime. There are precise, universal, mathematical, deterministic laws that govern the trajectories of test particles (having insignificant mass) in spacetime. Because of these similarities, the theories are described as *classical*; they will be discussed in chapter 3.

Quantum mechanics was developed over the course of the twentieth century by many scientists working in cooperation. It is far more difficult to characterise than classical theories because its physical meaning (its interpretation) is still hotly disputed. In quantum mechanics, time is more-or-less as characterised by Newton. In practice, in a good number of laboratory experiments, space can also be taken to be Newtonian. There have been attempts to combine quantum mechanics with relativity, as yet with partial success.

In quantum mechanics, there is a deterministic mathematical equation, the Schrödinger equation, which gives the dynamics of the entities within the universe. In contrast to the laws within classical theories, this equation is not quite universal: it ceases to hold whenever a measurement takes place. When this happens, the Schrödinger equation does not give the outcome of the measurement, but only encodes the probabilities that the outcome occurs in this location, or in that, or somewhere else. 'Measurement' and 'observer' are primitive, undefined concepts within the theory, but this strongly suggests that it cannot be the ultimate truth about the world: presumably the universe existed long before there were any observers around.

Quantum mechanics thus incorporates observers, and the measurements they make, into the universe, but only in a clunky way that is not explained. The system that is being investigated follows quantum rules, including the Schrödinger equation, whereas the observer and the experimental setup are described totally differently, using classical rules. But how can this be if the observers and apparatus are composed of particles that obey quantum rules? Some features of quantum mechanics are so bizarre that many prominent scientists have despaired of trying to understand the quantum world: it is forbidden to ask what is really going on; and a few deny that there is any quantum reality to understand. The theory, they say, is no more than a calculating device for predicting human experiences. The implication is that from now on physicists must give up their ambition of trying to understand the world. A more moderate view is that the quantum world exists, but experiments allow us to see it only partly, "as in a glass, darkly": we must continue the struggle to understand it. Roger Penrose takes this realistic and undespairing attitude (1989, pp. 346-48).

In recent decades, alternative interpretations of quantum mechanics have come to be more widely accepted and popular among physicists. These share certain general characteristics: they take a realistic view of the quantum world; give a uniform account of the universe as being governed by the same laws at all scales; they explain the properties and behaviour of large-scale objects (such as ourselves and laboratory equipment) in terms of their micro constituents (say particles); and finally they resolve (or at least attempt to resolve) many of the paradoxes of the theory in a clear, rational manner. Such interpretations do not take 'measurement' to be a basic concept. Indeed, supporters of the new interpretations attend conferences entitled "Quantum Mechanics without Observers".

Their lack of observers (at least none that are assumed as primitive elements) means that, despite their considerable theoretical attractions, such novel interpretations do not, of themselves, solve the mind-body problem. But we do indeed exist, at least in part, as observers; and any solution to the problem must explain this fact. I will show in Part III how one such novel interpretation can be adapted so as to provide an explanation.

Any attempt to sum up quantum mechanics in a few brief paragraphs is bound to oversimplify and contain contentious assertions. A fuller discussion of the theory will be given in chapters 4-6.

Objective physics

Current theories are highly successful in enabling scientists to make predictions with incredible precision, on scales ranging from much less than a billionth of a centimetre to that of the entire universe. Over a period of five centuries science, regarded as a human endeavour, has been highly effective in devising theories of ever widening scope and greater accuracy. Do these cultural achievements have their basis in some aspect of the universe's concrete reality, or are they merely down to luck? My view is that the success of science is grounded upon the existence of an objective physics belonging to the universe itself – an actuality that does not depend on observations made by humans (or by any other intelligence) – or even on their existence. Such a perfect, true physics I will call *objective physics*.

The concept of objective physics is not a scientific one because there is no experimental test that can determine whether current physics is identical to it. This was seen for example at the end of the nineteenth century, when Lord Kelvin asserted that the physics of his day was essentially complete, on the grounds that no experimental result was then known which was in conflict with the theories of his time. Very soon afterwards, new experiments and observations were made, which proved that these theories had serious limitations and which necessitated the wholly novel quantum mechanics.

Nowadays, as we have seen, our two best current theories are in conflict with one another, so we know that our current physics cannot possibly be the supposed objective one. Moreover, if objective physics indeed exists, this would imply, at least in principle, that quantum physics and general relativity can be merged into a new theory that encompasses both. However, even should such a unified scheme be discovered, it would still be open to the possibility of refutation, just as in the case of the science of Lord Kelvin's day. The concept of an objective physics is not part of science, but instead belongs to philosophy, which will be the topic of chapter 2.

The standard term for objective physics is *completed physics*. The latter expression is extremely misleading because it strongly and incorrectly suggests that intelligent endeavour is relevant to the concept. But one can readily conceive of a universe that remains lifeless throughout its entire history, which still possesses an objective physics. For all we know, our own universe might have turned out this way if its initial conditions had been slightly different. For this reason, I will always use the term objective physics.

A universe that did not possess an objective physics would be a horrible, chaotic mess. (Looking ahead, this would be the case for our universe if the Copenhagen interpretation of quantum mechanics is the literal and ultimate truth.) Throughout the book I will be assuming that our universe has an objective physics.

Issues in the vicinity of consciousness

As Chalmers points out, there are a number of issues of human functionality that are in the vicinity of consciousness (1996, pp. 26-28). My own, slightly different, list includes:

- The ability to use language
- The ability to integrate information coming from the senses
- The ability to store and recall memories
- The ability to make an internal representation of one's surroundings
- The ability act appropriately in accordance with one's memories and current representation

It is uncontroversial that all of these abilities can be realised physically, without reference to consciousness. Computer-based robots have been constructed that demonstrate all these abilities, albeit to a greater or lesser extent.

The question of whether or not such robots are actually conscious, or are merely simulating consciousness, will be left until later. As humans, when we recall a memory, this is sometimes, for example when we reminisce, a conscious experience. In contrast, computer recollection does not involve consciousness – at least not in any robot built to-date. The raw ability to integrate information is sometimes called *access consciousness*: by definition, access consciousness does not, of itself, involve consciousness – it is just a particular kind of information processing (and an unfortunate choice of words). In contrast, genuine consciousness (as I defined it above) is sometimes called *phenomenal consciousness*. To-date, robots possess limited access consciousness but no phenomenal consciousness. Would future robots, presumably solely on the grounds of increased complexity when integrating information, acquire consciousness in addition to access consciousness? This seems extremely dubious: there is no justification for anticipating that this could happen; moreover, if it did so, it would make zero difference to the robot's behaviour. In contrast, humans are frequently conscious when integrating information. The terms phenomenal and access consciousness were coined by Ned Block (Block and others, 1997, pp. 375-415), in which he also gave extensive examples of phenomenal consciousness in the absence of access consciousness.

The Hard Problem

Chalmers called the problem of giving a rational account of (phenomenal) consciousness *The Hard Problem* (1996, pp. xi-xiv). This problem challenges us to try to understand **consciousness** in the core, natural meaning of that word. Science describes the mathematical structure of the world, together with mathematical laws giving its dynamics. But nothing in science explains experiences, such as those of joy, or the sweetness of sugar. (Psychology accepts experiences as facts, but does not attempt to explain them.)

The problems in the vicinity, listed above, are, by comparison, *Easy Problems*. This is not because they are trivial – far from it. But with them, at least we have an inkling of how to proceed, and can begin to gain knowledge using the standard techniques of science: studying the brain, either anatomically or by using scanners; making cognitive models and simulating these on a computer; performing psychological experiments; and so on. The answers to these problems, while scientifically valuable in their own terms, do not touch upon consciousness proper – at least not directly (despite the phrase 'access consciousness'). Once the Hard Problem is solved, however, they will be invaluable in filling in the specifics of how consciousness relates to biological structures within the brain.

The purpose of this book is to make progress towards solving the Hard Problem. I believe that a rational solution exists, one which is consistent with current physics, but which itself lies beyond the scope of science.

Physicalism

Since the 1930s, and continuing up to the present day, one general approach to the mind-body problem has been dominant – accounting for over 98 per cent of all papers written on the subject. It exists in numerous varieties. This approach, called *physicalism,* takes the objects, properties, and laws of the physical world to be the fundamental things that exist. We are familiar with these things to the (excellent) extent that they are described by present-day science. According to physicalists, other things do exist that are not basic, for example 'nationality' and 'ownership', but these must have their full explanation in terms of the fundamental, physical entities. This reduction of everything to physics is reasonable for the two examples given, although admittedly the explanations will be long and convoluted.

A principle that all physicalists accept is the *causal closure of the physical world*: Physical events, insofar as they have a prior cause, are entirely caused by physical events. This principle, often abbreviated to *causal closure*, allows for the possibility of both deterministic and genuinely random events. (Here *genuinely* means that the randomness exists in the objective physics of the universe, and isn't just a reflection of our lack of knowledge.)

For the remainder of this chapter we will assume that physicalism is true; only at the end will I call this assumption into question.

Going back to Figure 1.1, **according to physicalists** the physical world (**C**, **D**) is fundamental and given. Under this assumption, the mind-body problem becomes that of giving a complete explanation of all mental facts (**A**, **B** and also perception, free will and agency) in terms of the physical facts. How successful has this approach been? Here – recalling the facets of the problem listed earlier – I will give a summary, which will be expanded upon:

1. *At first sight there are two, contradictory worlds: the physical and the mental*

1.1. *Which world are we living in?*

We are living in a physical universe (this is the definition of physicalism).

1.2. *What is the relationship between a person's mental state and their brain state?*

A person's mental state is fully determined by their brain state. As yet, there isn't a physicalist theory that expresses which brain state determines a particular mental state.

2. *How can minds perceive the physical world?*

2.1. *How do structural physical facts (a book is in front of you) become structural mental facts (your visual percept is of a book)?*

Neural activity in the brain must somehow reflect the structure of facts in the world. This problem remains unsolved, although there are no principled difficulties in finding a solution. What remains unexplained is, how does it come about that this neural activity is experienced?

2.2. *How can we explain qualia?*

No widely accepted physicalist account of qualia has ever been given.

3. *How can minds, as agents, cause changes in the physical world?*

3.1. *How are mental and physical causation related?*

The brain state underlying the current mental state is the sole cause of subsequent brain states and bodily actions. We can explain mental causation only if we identify mental states with underlying brain states.

3.2. *Can we make free choices? If so, how?*

Free choices must similarly be identified with underlying (deterministic or random) physical processes. According to physicalism, although we can have agency if we make this identity, we cannot have free will – at least, not as this is conceived of in the simplistic picture of Figure 1.1.

4. *How did consciousness evolve in a universe that was initially lifeless for billions of years?*

The short answer is that we cannot account for the emergence of consciousness in a universe that existed for many years without it. (This motivates the idea that consciousness is fundamental, existing everywhere and at all times: see proposal ***(F)*** below.)

Specific physicalist solutions to the mind-body problem

The physicalist mind-body problem asks: How are we to give an account of human (and other animal) minds in physical terms – i.e., in terms of spatiotemporal goings-on? Many differing specific answers have been proposed within the physicalist framework. A representative selection will be discussed, taking them **in order of increasing realism about mind**:

(A) Mind does not exist

This purports to solve the mind-body problem by asserting that the universe is nothing over and above a succession of physical events: → P → P → P →. Patricia and Paul Churchland, who describe themselves as "eliminative materialists", are the foremost advocates of this view. They claim that all psychological terms are nonsensical and naive 'folk' notions that will come to be replaced by scientifically rigorous objective concepts. Literal belief in experiential qualities, such as the greenness and scent of new-mown grass, will come to be likened to belief in fairies.

Can mind be eliminated in this way? One might be tempted to use Occam's razor to shave away the mind. The razor is often expressed as "Entities are not to be multiplied without necessity." But mind **is** a necessity: qualia for example are indisputable facts of our existence (despite the denials of some academics) and, as Russell affirmed, we are immediately acquainted with our experiences: an excruciating pain remains a fact whether or not it corresponds to a known bodily injury.

The major motivation for the denial of qualia is the widely held belief among academics that physicalism must be true because it is the only system consistent with science; and that one therefore has no choice but to deny qualia, and hence mind. We shall see in Part III that this belief is false.

Psychological experiments do undermine some of our naive intuitions about our experiences; and it has been shown that introspection often misleads us. But although we may be mistaken about many of our experiences, it is not credible that we are so comprehensively deluded that we are not having any experiences at all. Our experiences and our ability to act on them are crucial to our survival amid the dangers of the world. Moreover, scientists are fellow humans, and their sole evidence about the universe is, like ours, rooted in experience. All scientific theories are ultimately grounded in this manner, and so experience must be generally reliable.

Daniel Dennett's influential position is very similar to that of the Churchlands, though less directly expressed. In *Consciousness Explained* (1991) he gives many examples that cast doubt on the accuracy of human introspection, and by implication suggests that all introspection is worthless. He regards *phenomenology*, the study of experience, to be bogus and unscientific. Instead, he coined the term *heterophenomenology* (pp. 72-81) to denote the objective scientific study of subjects' verbal reports. His position is that the true meaning of the word 'consciousness' is what was earlier called access consciousness. In contrast, he regards phenomenal consciousness, phenomenology, qualia and the like as delusions. For this reason, he would regard a solution to all of the issues in the vicinity of consciousness to be a complete explanation of consciousness – period.

Dennett's position distorts the natural meanings of words: he accuses those who speak of blue in terms of experience as being deluded. Galen Strawson writes that Dennett is *looking-glassing* the notion of consciousness – referring to the famous quote: "'When I use a word,' Humpty Dumpty said in a rather scornful tone, 'it means just what I choose it to mean – neither more nor less.'" He is justly scathing about Dennett's bait-and-switch:

> I think we should feel very sober, and a little afraid, at the power of human credulity, the capacity of human minds to be gripped by theory, by faith. For this particular denial is the strangest that has ever happened in the whole history of human thought, not just the whole history of philosophy. It falls, unfortunately, to philosophy, not religion, to reveal the greatest woo-woo of the human mind. I find this grievous [...] (Strawson, 2006a, pp. 5-6)

(B) Illusionism

This is the theory that mind does exist, but it is nothing other than an illusion. The position is flawed: 'illusion' is a mental concept and so the notion that mind is an illusion is incoherent. One might try to get round this by redefining 'illusion' in solely physical terms, but this would be a case of looking-glassing the word, and this is subject to the objections discussed above.

An illusionist would claim that when we are seeing red, we only seem to be having this experience. But it is in the nature of an experience that seeming to have it and actually having it are one and the same. To give another example, if it seems to me that I am in excruciating pain, then I am indeed in excruciating pain. It may be the case that doctors have correctly diagnosed me as having a somatization disorder ('hysteria'); even so, the pain I am now suffering is real in the sense that it is an experience which I am now undergoing.

Another problem is that our knowledge of the universe comes to us solely through our experiences. But if mind is merely an illusion, why isn't the universe also entirely illusory?

Dennett is not an illusionist as defined here. Although he agrees that we seem to have qualitative experiences, and moreover that this illusion is often overwhelmingly convincing, he repudiates the idea that mind has any reality – even as an illusion. For him, illusions have no reality whatsoever – the universe is nothing but physics.

(C) Epiphenomenalism

Epiphenomenalism is the position in which (at least in some animal brains but perhaps elsewhere) physical states P cause distinct conscious mental states M, but conscious states have no effect on the physical world. The mental state *thirst*, for example, is distinct from and is caused by an underlying physical brain state (or dynamic) which we will call **thirst**. So, *thirst* is a feeling, whereas **thirst** is the particular pattern of neural firings which causes it. I do not reach out for a glass of water because of my feeling of *thirst*: it is the underlying neural **thirst** that causes my arm to move.

Strictly speaking, epiphenomenalism does not belong in the category of physicalism because mental states are understood to be distinct from physical states. It is loosely considered to be so here, because, according to epiphenomenalism, the physical world advances from one physical state to the next, taking exactly the same path as it would according to the eliminative materialism described above. Crudely, mind is just a garnish to the physical universe, playing no role in its functioning. There are many strong objections to epiphenomenalism:

First: *How mental phenomena are caused in the brain remains unexplained.* A neurobiological description of the state of a brain would presumably include: a description of the spatial configuration of neurons and how they are firing; a description of chemical transport and reactions, and so on. This same brain state is supposed to have subjective, qualitative and intentional properties or features caused by these objective neurobiological facts. How is a qualitative *thirst* caused by its neural correlate **thirst**? The theory is silent on this crucial matter. It is true that this question is hugely difficult – even to begin to answer; but it is a question that any qualia-realist is duty-bound to attempt.

Second: *Because it does not act on the world, mind can confer no evolutionary advantage.* This is contrary to what evolutionary biologists hold. Moreover, if mind is epiphenomenal, *how can we possibly practise science*: developing ideas, setting up experiments to test those ideas, expressing our ideas and the observed outcomes of experiments in reports?

Third: Here is my own, I believe decisive, argument against epiphenomenalism. Under this theory you can regard yourself as either essentially a physical or an experiential being. If you regard yourself as a physical being, then you can cut experiences away from the physical world by Occam's razor. You can function as a human organism, perhaps sometimes being in a physical state of **thirst**, without any need or use for *thirst*. This is essentially Dennett's argument (1991, pp. 403-405).

Alternatively, you may regard yourself as an essentially experiential being. Now epiphenomenalism leads to scepticism: Scepticism about other minds, because other minds have no effect on the physical world. (For the same reason, scepticism about your own mind as it existed in the past.) Scepticism about the physical world apart from the present moment, because the only thing that causes your present experiential state is the physical world at the present moment. Having got this far, isn't it plausible that the latter doesn't exist? Could it not be that the universe amounts to nothing more than your current moment of experience?

Epiphenomenalism therefore leads either to Dennett's behaviourist account, or to a scepticism which concludes that the entire universe consists of nothing except one's present moment of experience. Everyone finds the latter position absurd (it is called *Cartesian solipsism of the present moment*; named after Descartes, who introduced thoroughgoing scepticism as a useful tool in philosophy).

(D) Identity theories

According to *identity theories*, mind is identical to something physical (perhaps something functional). The great advantage claimed for this position appears to be that it seems to allow consciousness to play a causal role in the physical world. We shall see below that this initial hope is false. David Papineau is an articulate advocate of identity theory, and he stated it cogently in a YouTube debate with non-physicalist Philip Goff (Goff & Papineau, 1967). I will try to summarise his position here as accurately as I can, while omitting certain technicalities.

As a preliminary remark, I must mention that *identity* is always understood in the strongest possible sense. It isn't just that things are as alike as two peas in a pod, or as two copies of the same book, or like identical twins. A and B are by definition *identical* only if they are **one and the same**. In the sense used here, I'm not identical to my brother David, even though we are twins. But I am identical to the person writing this sentence.

Papineau agrees that we possess minds, phenomenal consciousness, percepts, qualia and all the rest. He also agrees that we are perfectly acquainted with many aspects of our experiences: when we feel we are in pain, there can be no doubt that we are in pain.

Papineau holds the position of *conceptual dualism*: we have two concepts, that of a percept of red, and that of a certain neural situation. At first sight, these seem radically different, but despite this, they both refer to one physical condition: a certain type of pattern of neural firings. For example, a doctor may be monitoring your visual cortex in a scanner, while you are looking at a red patch. The physical goings-on, partially revealed by the scanner, and in addition by your visual experience, are both aspects of one thing: the true and objective physical pattern of neural firings going on at this moment in your brain.

How can these things, so seemingly disparate, be identical? Papineau's answer is that whilst we have infallible knowledge of some aspects of our experience (such as knowledge of when we are in pain), our knowledge of other aspects is fallible and incomplete (we do not recognise that our experience is identically a pattern of neural firings). Papineau agrees that this is surprising and counterintuitive, but argues that it is nonetheless true, and we should just get used to it. He makes the analogy that we do not experience the fact that the earth is a sphere in rapid motion, but this is the case nonetheless.

The argument could also be made (although Papineau doesn't mention this) that our ignorance of the fact that our experiences are identically neural firings is evolutionarily essential for our survival: If we were conscious of our neural firings, in addition to our percept of the tiger, then this would be a huge and dangerous distraction.

Having described the points in its favour, I turn to critiquing identity theory.

First, Papineau's theory explains why, **in our day-to-day lives**, we do not recognise experiential states as being identical to brain states. But this is not sufficient: What is also needed – as a prerequisite for this explanation to go through – is **a physical account** as to why complex spatiotemporal processes in the brain are accompanied by technicolour visual experiences, olfactory sensations, pleasures and pains – in short, by qualia. Such an account is necessary because identity theory is supposedly a form of physicalism. As yet, however, no identity theorist has supplied the crucial justification as to why we have any experiences in the first place. Identity theory thus suffers from what Joseph Levine calls an *explanatory gap* between the seemingly incompatible natures of our subjective experiences and neural firings (Levine, 1983). This intractable difficulty applies more generally – to any physicalist account that holds qualia to be real.

Second, there is the problem of lack of true agency (which applies to all physicalist theories). Physicalists take the sole form of causation in the world to be physical; either deterministic or genuinely random. A human brain is no exception to this. Considered as a physical system, the brain goes from one state to the next, according to the laws of objective physics. The fact that the 'owner' of this brain has certain percepts, feelings of striving and of volition makes not one jot of difference to the sequence of physical events. How then do persons truly have agency, let alone free agency, under the assumption of physicalism?

There is a huge amount of literature on the subject of free will. Usually it is studied in its own right – quite independently of the mind-body problem. Because most scholars are physicalists, the usual approach is that of *compatibilist free will*. Two tenets of this position are: (1) A person acts according to their desires, but there is no way that they can control their desires; (2) If that person's brain state had been a little different, then they might have acted differently, according to their different desires.

According to compatibilists, (1) amounts to agency, and (2) amounts to free agency. Both of these claims are tendentious: In (1) a person's brain is just following the laws of physics. Their desires are wholly irrelevant, and so this cannot amount to agency. Proposal (2) refers to a fictional brain state, different from the one that the person actually has. What possible relevance could this non-existent brain state have to the person's freedom?

When Bernard Lovell decided to build a radio telescope to study the heavens, the result was the creation of a 76 metre diameter dish that could be pointed very precisely to any part of the sky. Lovell's choice resulted in huge physical changes to the landscape (in its current form the instrument weighs 4.7 million kilograms); more importantly, his decision set the overall direction of the whole future course of his life.

If humans possess free will, then the above paragraph is plausible, rational, and factual. On the other hand, if we merely have compatibilist free will, then the above events are due solely to: the initial state of the universe; the laws of objective physics; and perhaps to objective randomness (if this is a feature of the laws). Human subjectivity is entirely irrelevant. This is not only wildly implausible, especially when decisions are significant, as in the Lovell example, but also has the many absurdities listed under ***(C) Epiphenomenalism***.

If physicalism is true, then my protest above is irrelevant: the consequent causal closure of the physical world makes compatibilism inevitable. **But physicalism might be false.** If this is so, then alternative accounts of free will become available. A detailed discussion of free will in physicalist and non-physicalist contexts is postponed until chapter 9.

Third, there is no possible way that one can determine, on the basis of a creature's (or other entity's) behaviour, whether or not it is having experiences. Identity theorists generally assume that we can establish this: but it clearly follows from the previous objection that this is not the case.

There are further objections to identity theory, but those listed above are sufficient to nullify it as an adequate account of the mind-body problem.

(E) Emergence

Mind and qualia exist in their own right because they are 'emergent'. The overwhelming consensus of scientific and philosophical opinion is that for billions of years the universe existed as a physical system, without the least smidgen of consciousness. This late arrival must have emerged from, and have been caused by, earlier goings-on, which have a satisfactory explanation in purely physical terms. Let us look at this consensus in increasing detail.

In cosmology, following the Big Bang, change is for the most part gradual, and the same physical laws apply everywhere and at all times. There is thus continuity between the very early universe and the cosmos as it is now. The former is generally assumed to have been absent of any experience whereas, in the present epoch, consciousness exists on this planet, and plausibly on others. Imagine colouring in a space-and-time map to show that portion of the universe where, according to the consensus, consciousness exists or has existed. This map would be blank until recent cosmic times; and even today, at best there would be some sparsely scattered specks of colour. A solution to the mind-body problem would either have to provide some account of this meagre distribution of consciousness (including a principled account of its boundaries), or explain why the consensus is wrong.

In evolutionary biology, new species develop very gradually

by means of natural selection over the course of ages. In a single generation, offspring are very similar to their parents. It follows that: either, at some point in the Earth's history, an animal entirely lacking consciousness gave birth to an offspring possessing some minuscule sensation (perhaps a tickle); or alternatively, every animal throughout the history of life must have had at least a trace of experience.

The argument within human developmental biology follows the same lines. There are smoothly changing physical processes controlling the development from a fertilised egg, to a foetus, to a baby, and to a human adult; all underpinned by the same physical laws. The consensus view is that a fertilised egg functions as a physical system in the absence of experience. At some point along the human journey from egg to adult, consciousness must appear – in primitive form at first. A solution to the mind-body problem must give a principled account of this boundary, and explain what difference the possession of consciousness makes.

There are two objections to emergentism. First, *mild emergentism*: This proposes that mind arises when certain physical conditions are satisfied. No one working within this model has, as yet, given a principled account of where the boundaries between regions of non-experience and experience lie. What **specific** physical conditions hold within the tiny portions of spacetime where experience exists? And why do these conditions entail that there is experience? One proposal – attractive only at first sight – is that consciousness arises in the brains of creatures possessing central nervous systems. But this suggests there is something mysterious and special about the physics of neurons – and this is not the case. Adding criteria about complexity or functionality do not help to clarify matters here.

Second, *radical emergentism*: This proposes that a novelty arises which is not a logical consequence of the objective physics of the world. This clearly goes against the spirit of physicalism.

It might be possible to reject physicalism by asserting that the novelty is non-physical, but I know of no one who has taken the route – the onus would be on them to give a rational account of this novelty.

My view is that, as so far developed, 'emergence' is little more than a comforting word that does not amount to an explanation. Philip Clayton gives a survey and defence of emergentism (2004).

(F) Panpsychism

Panpsychism is the claim that mind exists everywhere in the universe. Even the ultimate entities of objective physics (electrons are possible examples) have mind or mind-like properties. At first sight this seems wildly implausible but it is motivated by the complete failure of emergence theories to give a rational account of the supposed boundary between regions of the universe possessing experience and those without. Panpsychism solves this problem by asserting that there are no such boundaries, and there is no need for emergence, because mind is omnipresent right from the universe's beginning.

Panpsychism also goes partway to solving the problems of qualia: these are taken to be fundamental, and so, of themselves, do not require an explanation. However, an account is still required as to the causal role that qualia play in the dynamics of the world, and how they relate to traditional physical properties, like mass or charge. (The difficulty is that physical properties alone seem sufficient to explain the dynamics.) The causal role of qualia is a tough problem that panpsychists approach in different ways, with varying degrees of success.

Another argument in its favour of panpsychism is that we generally suppose that experience exists in at least those creatures that are biologically close to us. Most of us find it plausible that certain creatures biologically simpler than us, say birds, might have certain limited experiences, even though we cannot yet prove this. It is rational to hold that experience may be more widespread than we currently know. Having conceded this, it must at least be possible that mind is everywhere.

Panpsychism is not so very radical: it merely asserts that something, which at first sight we take to be distributed in a few places in the universe, is actually present everywhere. As such, the above proposed solutions, ***(B)*** through ***(D)***, to the mind-body problem, and their difficulties, also apply to panpsychism. (***(A)*** and ***(E)*** are excluded because they are the negation of panpsychism.) Options ***(B)*** and ***(C)*** remain wholly unacceptable for the reasons originally given. The conclusion is that, of the **physicalist** approaches which we are considering here – and I've tried to choose the best – ***(D) identity theories*** are the most credible basis for panpsychism.

(At present we are assuming physicalism is true. Pan-idealism, to be explained in Part III, can be regarded as a variety of panpsychism; but it is idealist rather than physicalist in its nature. As will be shown later, pan-idealism does not suffer from the problems of physicalist panpsychism.)

As with traditional identity theory, however, there remains the problem of explaining human free will (which was also described earlier under the name 'reflective agency'). As yet, this has received little attention from panpsychists – again revealing the widespread bias in which the mind-body problem is addressed mainly from the perspective of perception. One hope is that the ubiquity and fundamentality of bare experience and bare agency give panpsychists additional tools to tackle this problem – tools not available to identity theorists in general. Panpsychists can accept without qualms that the observed randomness of photons in certain experimental situations is a manifestation their bare agency.

Another, and certainly the best recognised and most difficult challenge facing panpsychists, is the *combination problem*. How do human beings, composed of myriads of individual atoms, each of which is a centre of experience and agency, come to have that unity of mind that we know we have?

David Skrbina has written an accessible survey of panpsychism (2003). Other important works are Brüntrup & Jaskolla (Eds., 2017) which has no fewer than five chapters devoted to the combination problem, and finally Seager (Ed., 2020), which has 28 chapters, each by an established expert.

Moving forward

Throughout this chapter I have been assuming physicalism. But physicalism does not lie within science – it is a philosophical position. Many scholars suppose that physicalism is the only viable position; they consider it to be the only possibility that both fully respects and fully engages with the best of current-day science. This is their motivation for clinging on to it – despite almost a century of intensive efforts, all of which have failed to solve the mind-body problem, and with no prospect of success in sight.

But, as we shall see in Part III, there are alternative philosophical systems that engage with physics extensively and successfully. Moreover, these systems give a much more satisfactory and reasonable solution to the mind-body problem. In preparation, the next chapter lays out the philosophical groundwork.

Chapter 2

Philosophy

We have seen that physicalism is a philosophical position which works wonderfully in almost all situations; but, in all its varieties, it has severe difficulties when it comes to tackling the mind-body problem. Given below is an account of those parts of philosophy most relevant to confronting this difficulty. (For instance, 'continental philosophy' will be omitted because typically it does not engage sufficiently with science.)

This chapter will discuss: the character of philosophy as a discipline; two major methods of carrying out philosophy (analytical philosophy and metaphysics); how to compare metaphysical systems; and criticism and defence of philosophy. Along the way some necessary philosophical concepts will be defined. The final section sketches, in very general terms, a particular metaphysical position, called *pan-idealism* (an alternative to physicalism) that will be developed and defended in Part III.

What is philosophy?

Philosophy comes from an ancient Greek word meaning 'love of wisdom'. It encompasses any method of rational enquiry into any topic. The scope of philosophy is thus huge: we can have the philosophy of X, where X is any noun. Examples include: philosophy of science, ethics, aesthetics, language, religion, mathematics, mind, Artificial Intelligence, or even philosophy of philosophy (*metaphilosophy*). This book lies within the domain of *philosophy of mind*.

Philosophy can be contrasted with *science*, which may be characterised as being concerned with *empirical* knowledge, that is to say, with knowledge that can be determined by experiment or observation. The vast scope and success of science is universally agreed and understood. Some things are outside the remit of science. As we saw in the last chapter, the contention that the universe has a true, *objective physics* is not part of science because it is not empirically testable; it is a philosophical claim. As questions in philosophy go, this claim is exceptionally watertight, and it will not be disputed in this book. Recall that the standard philosophers' term for objective physics is *completed physics*. I do not use this term because it wrongly suggests that the concept could eventually become part of science.

There are two contrasting general areas within philosophy: First, *ontology* – the study of existence. What are the characteristics of things that exist? What do we mean when we claim that something exists? Second, *epistemology* – the study of knowledge: What can we know with certainty, and on what rational principles do we base our knowledge?

There are two distinctive methods of going about work in philosophy. *Analytic philosophy* has been in vogue since the beginning of the twentieth century. It involves dialogue and rational argument from a few agreed premises. *Metaphysics* is the setting-up of an all-encompassing world system. An important part of a *metaphysical system* is its fundamental ontology. Next there is *causation* between these basic entities: this explains how goings-on in one fundamental entity makes changes happen to the state of another fundamental entity. Finally, there is the requirement to **explain non-basic entities and goings-on** in terms of fundamental entities and fundamental causation. For example, physicalism is a metaphysical system, its fundamental entities are those of objective physics, and its causation is provided by the laws of physics. Within physicalism, non-basic entities include biological entities (say birds), and economic systems. Metaphysics has declined in popularity over the last century. Both of these methods will be discussed more fully below.

Philosophy has been heavily criticised in recent years, on three main grounds. First, philosophy has seemingly made little progress over the centuries, ploughing over the same ground again and again. Second, the rapid rise of science is thought by some to have been superseded and encompassed by science. Those parts that cannot be assimilated are nonsense, it is alleged. Third, based on the previous points, it is claimed that rival philosophical positions cannot meaningfully be evaluated against one another. I will answer these objections toward the end of the chapter.

Analytic philosophy

For many decades, almost every philosopher has claimed to be an analytic philosopher. *Analytic philosophy* is a vague term, but broadly speaking it is a method of philosophising that models itself on logical or mathematical proof. Ideal examples of the latter are provided by proofs in Euclidean geometry. The method assumes that: We can define words exactly, "A *line* is a length without breadth." We can specify postulates that will be accepted by all reasonable people, "It is possible to draw a circle centred at any point, and with any radius"; and so on. Another requirement is that arguments are made with strict logical rigour.

In analytic philosophy, the postulates are taken from some default metaphysical position, call it P. The aim of an analytical argument is to refine this position, typically by clarifying the meanings of words. In philosophy, each logical argument is short – usually involving about half a dozen steps. Successive analytic arguments, it is hoped, will result in a highly polished version of P, adequate to provide accurate and convincing descriptions in the widest possible variety of situations.

For example, the metaphysical system is very often taken to be some variety of physicalism. One postulate might be "the causal closure of the physical world". A general idea from physicalism is that, given the exact physical state of a person's (or animal's) brain at a particular moment, their mental state is fully determined. The philosopher's jargon for this concept is that the mental state *supervenes* on the corresponding brain state. There have been many analytic arguments about the precise meaning of the concept of supervenience, leading to a great many, subtly-different definitions of the word.

It is important to argue correctly. A particular fault is called *begging the question*. In philosophy, this involves assuming the conclusion you are trying to reach in the course of making the argument. This is clearly a fault, but an easy one to make inadvertently. In everyday speech, this error is sometimes referred to as "making a circular argument". An example from philosophy is: "Physicalism must be true, because, in principle, physics will eventually tell us everything it is possible to know about the ultimate constituents of nature."

Somewhat confusingly, in everyday arguments nowadays, people more often use the expression "begs the question" to mean "raises the question" (which is not a fault). For example, "The government tells us that prices will fall. This begs the question, 'How far?'" The term is never used in this manner in philosophy.

Critique of analytic philosophy

- *Analytic philosophy has a built-in bias towards the default metaphysical position*

Analytic philosophers presume that some default metaphysical position is true; and they will continue to stand by this position until some watertight proof demonstrates that it is logically inconsistent. This means that analytic philosophy puts an inappropriately heavy burden of proof on those who wish to argue against the default position.

- *Analytic philosophers behave as though we can postulate an adequate number of solidly known facts*

But as Descartes, Hume and Locke have shown, what can be postulated with absolute certainty in philosophy is pitifully small. At best there are two or three unexceptionable postulates, because philosophical terms are theory-laden. There are several attitudes we might take towards this uncomfortable situation. We can:

1) Cling to these few crumbs.
2) Ignore these great philosophers and pretend that there are sufficient postulates that cannot be faulted.
3) Concede these facts, and instead enquire as to what may be reasonably claimed with overwhelming plausibility – rather than absolute certainty. (For example, we have great confidence that the sun will rise tomorrow.)

The first option allows for very few sound arguments, and this stops analytic philosophy in its tracks. The second option is disingenuous, but – because most arguments in analytic philosophy are inevitably grounded upon questionable postulates – this is implicitly what analytic philosophers are doing. (Although postulates might perhaps be questioned, it is impossible to do so without getting into a metaphysical debate, which most analytical philosophers deplore.) The third option is the only viable alternative.

- *Analytic philosophy requires inappropriate certainty in its rules of argument*

Scientists use the principle of scientific induction, which generalises evidence that is limited both in time and space, in order to discover universal laws. While induction has been spectacularly successful, it doesn't result in certainty.

It is strange that philosophy, which discusses the character of the world in very general terms, and which is not experimentally or observationally testable, should require logical certainty in its method of argument. Philosophers should be more modest in what they claim for their arguments: certainty is only available in two or three instances; most arguments are merely reasonable.

- *Analytic philosophy presumes that refinement of ideas is sufficient*

Because analytical philosophy is a process of refinement, and many philosophers use no other method, the implicit assumption is that this process is always sufficient: That there is never the need to investigate any alternative metaphysical systems, which may be wholly different.

The default metaphysical position is usually physicalism, but this ignores the possibility that physicalism is radically false. It may be possible to start from scratch and develop an entirely contrasting metaphysical system (which must of course remain consistent with current scientific knowledge). Such a metaphysical change would be somewhat analogous to a paradigm shift in science – as happened when Einstein's ideas superseded Newton's.

To sum up, analytic philosophy has its rightful place as an important and useful implement in the philosopher's toolbox, but it cannot be the only one. Although most present-day philosophers are chary about metaphysics, analytic philosophy is grounded in the latter method.

An excellent example of a discussion between analytic philosophers may be found here: (Goff & Papineau, 1967). It is interesting for the quality of the debate, the fact that they are discussing the mind-body problem, and that one of them, Philip Goff, is a non-physicalist.

Metaphysics

Metaphysics is notoriously difficult to define, but John Cottingham's statement is excellent:

> [Metaphysics is a] philosophical enquiry that goes beyond the particular sciences and asks very general questions about the nature of reality and the ultimate conceptual categories in terms of which we are to understand it.
> (Cottingham, 2008, p. 68)

Speculative metaphysics (Alfred North Whitehead's term) goes further than asking **questions**; it attempts to provide a rational, comprehensive **system** or model for understanding the cosmos both in its particulars and its entirety. A metaphysical system should have ambitions to be comprehensive – encompassing not just science but also ethics, aesthetics, mathematics, and so on. Of course, this ideal is extremely difficult to achieve.

For several centuries metaphysics has come under attack. Here are quotations from two eminent critics: David Hume (1748) stated that any book on metaphysics should be "committed to the flames, for it can contain nothing but sophistry and illusion"; and A.J. Ayer (1936, p. 55) claimed that, "Philosophy, as a genuine branch of knowledge, must be distinguished from metaphysics. [...] The majority of the 'great philosophers' of the past were not essentially metaphysicians." Ayer's final point is implausible: Around 400 BCE Democritus advocated the first atomic theory, without entertaining any hope that atoms might be detected. In 1641 Descartes proposed a two-substance theory in which mind is not physical – it is somewhat analogous to the simplistic picture of Figure 1.1. Both systems are most straightforwardly understood as being metaphysical. Criticisms of metaphysics boil down to the fact that metaphysical systems are, by definition, non-empirical: otherwise they would instead come within the ambit of science. Ayer and Hume claimed that such systems are therefore meaningless.

Monism is the belief that the universe is made up of one kind of entity or stuff. Physicalism is an example of a monistic theory. Note that the phrase "one kind" is allowed great leeway: although there are differences between charged and uncharged particles, and even greater differences between particles and fields, they are all, broadly speaking, spatiotemporal. In contrast, *idealism* is a type of monism in which the fundamental things that exist are all minds, or are mind-like in character. Idealism comes in many variants, and some will be discussed below. *Dualism* holds that there are two distinct kinds of fundamental entity: in Cartesian Dualism, these kinds are matter and mind.

In opposition to Ayer and Hume, I hope to show how metaphysical systems can be compared rationally despite the fact that they are non-empirical. As examples I will use three contrasting metaphysical systems: Berkeley's idealism, (standard) physicalism, and panpsychism.

Berkeley's idealism

In Berkeley's system, the primary or foundational things that exist are minds or souls. Objects that we usually think of as 'existing in the external spatiotemporal world', such as a football or a river – even our own bodies, and indeed the whole universe – also exist, but only as percepts, which are specific types of experiences within minds. All such objects are thus secondary in Berkeley's ontology, and for them he made the famous definition of existence: *esse est percipi* ('to be' is the same as 'to be perceived'). Berkeley also argued that mathematics exists solely as thoughts within minds (1710, introduction).

As thus far developed, Berkeley's metaphysics is a cosmos of disjointed minds, each with its own separate and unrelated contents. Something else is needed in order to bind this system together into one coherent whole. Berkeley therefore proposed that the Christian God – an all-powerful and beneficent mind –

places percepts (in a lawful manner) in the minds of individuals, in such a manner that they have logical and mutually consistent experiences: Water is a well-behaved liquid, consistently obeying the God-given laws of physics. Each person in a given situation has experiences consistent with those of the other persons present. Each member of a punting party, for example, can see their companions, talk to them and hear them, and so on. In Berkeley's system, the very concept of people **being co-present in a given place** is that their experiences are in harmony. God also ensures the continued existence of (say) a book when it is locked away from human observation in a drawer: it remains a percept in God's mind, and thus it continues to exist under Berkeley's definition.

People possess what philosophers call *libertarian free will*. A person can choose to raise their hand and it will rise; moreover a person can choose between conflicting desires, and is free to act according to this choice. In Berkeley's idealism, free will operates under God's direction in the manner already described: God places appropriate and consistent experiences in the minds of all persons present.

Considered in its own terms, Berkeley's universe does possess an objective physics. Physical entities, with their physical properties, exist in the mind of God, even when unseen by humans. Also, physical laws exist objectively as ideas in God's mind. (These laws are chosen by God.)

Berkeley's attack on matter

Berkeley's attack on the concept of matter, defined as an insentient substance that exists in an external spatiotemporal universe, was a critique of John Locke's position. Consider a putative material object – a desk. Locke argued that the properties of the desk could be divided into two categories: The qualitative colour brown of the desk, the qualitative scent of furniture polish and qualitative squeak of the drawer are all secondary qualities and exist only in the mind of the observer. The primary qualities of the desk are its spatiotemporal extension, form and motion. These are all characterised by being measurable, and only these, said Locke, truly exist in the external world. (Locke also included solidity as a primary quality, but this can be defined straightforwardly in terms of spatiotemporal behaviour.)

The true physical character of the desk is thus somewhat akin to an animated (e.g., its drawers can be moved) mathematical figure. Locke argued that the difference between a real desk and a mathematical object was that with the real desk there was an underlying substance called *matter*, which acted as a "support" that bound its primary properties together into a whole. Berkeley rightly and roundly criticised this obscure Scholastic idea on the grounds that this substance was wholly unobservable and nothing could be known about it. Moreover the metaphor of "support" has no clear meaning: it cannot be "as the legs support the body" (Berkeley, 1713).

Berkeley questioned the sharp distinction between primary and secondary qualities, maintaining that: First, both types of quality are subjective: just as the desk would look grey in the twilight, so it would have a different shape when viewed from a different angle. Second, primary and secondary qualities are inextricably bound together. The rectangular shape of the desktop is only revealed to us because we experience it as a trapezoidal patch of colour, which contrasts with the colour of its background. Berkeley went on to argue that, since secondary qualities undoubtedly exist in the mind, then so must the primary qualities. Here is some of his vivid prose on this point, "all the choir of heaven and furniture of the earth, in a word all those bodies which compose the mighty frame of the world, have not any substance without a mind" (1710, paragraph 6). In this statement "without" can be taken in the modern sense of 'in the absence of', or in the older sense of 'outside'.

Physicalism

Physicalism has already been met with in chapter 1. It is by far the most popular metaphysical position of our time. Many persons assume physicalism by default, believing it to be synonymous with science. It exists in many variants, but here are some of its most important and commonly-held theses, compressed from Poland (1994):

- Everything that exists in nature is dependent upon the physical domain
- In any part of the world, the physical facts determine all the facts
- Different branches of knowledge are organised hierarchically with physics at the foundation
- Causal closure: for every physical event that has a cause, this cause is physical

According to physicalism, at the foundational level we have physics, with particles, fields and so on, describable by highly-complex mathematical laws. Given these, and a set of initial conditions, we have a universe, the beginning of whose evolutionary history is accounted for solely in physical terms. As the universe cooled, it began to exhibit *chemistry*, and stars and galaxies formed. On at least one planet, life began to evolve into ever greater complexity, and so *biology* arose. Then something extraordinary happened: *mind* arrived in the universe. At least some creatures began to have qualitative experiences of their surroundings. Dogs can experience the colour, taste and scent of meat, hear sounds, and so on. Human beings can in addition reflect upon their experiences. According to most physicalists, mind was a recent arrival in an infinitesimal part of the universe.

As can be seen, physicalism attempts to stay close to current scientific knowledge. But physicalism is not identical to science:

it relies not only on the bare facts, but on the metaphysical principles just listed.

Problems of physicalism

(I) *The explanatory gap.* This has already been discussed in chapter 1: There is "something it is like" for us to see a rose: we have a qualitative experience of (say) yellow. How could the spatiotemporal behaviour of systems of matter in either the rose or in the brain (regardless of their complexity) necessitate the existence of / give rise to / be identical to the 'yellowness' of the experience of seeing the rose? Similarly, how could physical goings-on in the brain be related, in any of these three ways, to the pain of a migraine?

It is incumbent upon physicalists to give an explanation **in physical terms** as to why we have such experiences. This is because to be a *physicalist* is, by definition, to make the claim that physical goings-on in our brains necessitate the fact that we have experiences. As Colin McGinn puts it:

> How is it possible for conscious states to depend upon brain states? How can technicolor phenomenology arise from soggy gray matter? What makes the bodily organ we call the brain so radically different from other organs, say the kidneys – the body parts without a trace of consciousness? (McGinn, in Block, Flanagan & Güzeldere, 1997, p. 529)

For completeness, it is worth adding that McGinn is a *mysterian*, meaning that he remains a physicalist, but argues that, because of our innate biological limitations, we do not possess the intellectual capacity to solve the mind-body problem. Of course, this argument does not itself amount to a solution of the mind-body problem.

(II) *What do we mean by the term "physical"?* This is sometimes called *Hempel's Dilemma*: If we mean present-day physics, then physicalism is trivially false. If we mean the (perhaps vastly different) objective physics of the universe, then physicalism is hopelessly vague. For example, we might attempt to 'solve' the mind-body problem by asserting that the objective physics of the universe includes qualia. This promissory note predicts a wholesale change in the character of science; even so, it does not of itself amount to a solution.

(III) *Updating Berkeley's attack on 'matter'.* Berkeley rejected general abstract ideas, but (putting aside the controversial issue of what it means for a mathematical object to exist), an argument similar to his still goes through: How does an unobserved physical system differ from a mathematical model of it? The unobserved physical system might be a portion of the microphysical world, the very early universe, or a mouse hidden in a drawer. As I will argue in chapter 7, physicalists make no distinction between a real (instantiated) entity, and a corresponding merely putative (uninstantiated) entity. Both are characterised solely by a physical description which is entirely mathematical.

This is fatal to physicalism. For example, a man is alone in a room, watching television; he is attending to the show and not observing himself. Consider both the actuality of this situation, and the completest possible mathematical description of it in terms of objective physics. In physicalism there is no distinction between these two things. Because of this lack, any proof (or explanation) that mind arises in the man's brain can be translated into a proof (or explanation) that mind must arise within the corresponding mathematical model. We can conclude that: either no proof exists that mind necessarily arises, in which case mind must forever be a mystery to physicalism; or, alternatively, mind necessarily arises from within complex mathematical systems, which is absurd.

(IV) *The problem of free will.* Physicalism is inconsistent with libertarian free will. Because of causal closure, the only form of free will that a human being might have is compatibilist free will, as defined in chapter 1. This is not at all plausible for significant human choices. Moreover, scientists have to be free to choose which experiments they perform, and to make their own judgements as to what conclusions they draw from experimental results.

(V) *The boundary problem.* How is one to specify **in physical terms**, the boundary between experiencing and non-experiencing regions of the universe? Moreover, does one rationally justify this specification?

Criteria for comparing metaphysical systems

Metaphysical systems are not empirical, but we may use criteria such as those listed below, in combination, to help us make reasoned judgements between them.

One way of comparing metaphysical systems is in terms of their *scope*. This is the range and variety of facts adequately explained by the system. Other things being equal, a metaphysical system with broader scope is better than a narrower one. But caution is required: If the putative explanations concerning a certain domain of knowledge are wholly inadequate, then the scope of the system cannot honestly be said to encompass this domain. This would be the case, for example, if a metaphysical system's putative account of the mind-body problem turned out to be entirely deficient. All the systems being considered here are identical in their **claimed** scope, because they seek to give accounts of two important domains: that of the mind-body problem, and that of the physical domain. Adequacy in each of these domains are the first two criteria in the list:

(1) ***Adequacy*** *in giving an account of the mind-body problem*

This is the key focus of the book.

(2) ***Science****: Engagement with and consistency with current science*

This is essential of course. But there is a crucial word of warning here: Current science consists in what can be determined *empirically*, that is to say by experiment, observation, or by the mathematical theories arising out of these. But beware that science **does not include anything that more properly lies within the ambit of the philosophy**. For example, the statement that "all mental states must be grounded in physical states" is a physicalist claim that is not empirically testable; and so (whether true or false or meaningless) it does not belong to science. This criterion can be regarded as a special case of the next:

(3) ***Accepting basic facts*** (i.e., not denying them)

A metaphysical system that makes the claim that "Time does not exist", for example, would have to provide solid reasons for this.

(4) ***Plausibility****: This should be measured from within the system itself, considered in its entirety*

In contrast to this procedure, suppose we have two rival systems A and B, where A is currently generally accepted, and B is a minority position. It is not a fair comparison to assume A, and then claim that B is implausible because one of its postulates is incompatible with those of A (or their consequences).

(5) ***Minimal brute facts****: The number of brute facts or miracles needed should be minimal*

A *brute fact* is a basic fact about the world that has no explanation, and must be assumed true. It is akin to an axiom in geometry. Some postulates are necessary in any metaphysical system. But once our universe is 'up and running' so to speak, then brute facts and miracles are equally undesirable. At one time, for example, the solar system was thought to be unstable, and Newton argued that God occasionally nudged the planets to correct this. Clearly this is a weak idea; but to pretend to solve the difficulty by adding "Solar systems are stable" as an extra postulate of Newtonian physics would be equally undesirable.

(6) ***Minimal promissory notes***

These are assertions of the type: "Science, in the far future, and using utterly different methods, will solve this problem eventually."

(7) ***Elegance and simplicity***

For example, Descartes' two-substance theory, in which mind and matter are distinct types of entity, is less elegant and much more complex than physicalism, in which the only kind of substance is physical stuff. Causation within Cartesian Dualism is also complicated. There are four distinct types of causation: body-body (also called physical-physical); body-mind (perception); mind-mind (train of experience); and mind-body (agency). Explaining how these four operate in harmony is challenging. Of course, if no simple system adequately explains the facts, then a more complex metaphysic may have to be accepted.

(8) ***Clarity versus vagueness***

For example, emergentist theories tend towards vagueness. An eye begins its evolution as a pit of cells 'sensitive' to light. In a simple creature, this sensitivity allows it to behave in an appropriate fashion, either seeking out or hiding from the light. The human eye/brain system is also 'sensitive' to light. In the first instance this sensitivity is no more than a capacity to behave appropriately. In the second case, sensitivity involves perception and qualia. Somewhere along this more-or-less continuous evolutionary pathway, the meaning of the word 'sensitive' has inexplicably changed radically. The primary focus for emergentists – by the very nature of their position – should be upon **(V)**, *the boundary problem*. Yet very few emergentists acknowledge, let alone tackle it.

(9) ***Naturalism****: a metaphysical system should not involve religious ideas*

Even among the minority of philosophers having a religious faith, I think most would assent to this.

Some might wish to choose a somewhat different set of criteria; but in this situation one would expect there to be a good deal of overlap. In any event, despite possible disagreement, the above criteria provide a sufficient basis for philosophers to have fruitful dialogues in metaphysics, to improve their existing systems, and to devise wholly novel and better systems.

The next task is to evaluate the metaphysical positions discussed here against these criteria. Depending on the metaphysical system, the criteria must be evaluated in a particular order, because they sometimes interact.

Evaluating Berkeley's idealism

In my view, Berkeley scores well on criteria: (3) **Accepting basic facts**; (6) **Minimal promissory notes**; and quite well on (7) **Elegance and simplicity**. Many people come down hard on point (4), the **plausibility** of his system. I believe that this is unfair, because the system, considered in its entirety, straightforwardly makes sense: it is a homogeneous world consisting of many finite minds, and one infinite mind that plays an essential causal and unifying role. Berkeley's definition of existence is appropriate within his system.

On criterion (8) **Clarity versus vagueness**, Berkeley is somewhat vague as to whether higher animals have minds. A greater fault is at criterion (5) **Minimal brute facts**, because every fact about the universe is a direct act of God's will – a miracle. The greatest weakness of Berkeley's idealism comes at criterion (2) **Science** because, although there is consistency with science, there is a total lack of engagement with the specific facts of science. It is entirely arbitrary: whatever the facts turn out to be, Berkeley simply asserts, in effect, "God does it that way." Berkeley's idealism also fits badly with the well-attested facts of biological and cosmic evolution: it is a very deceptive God who would plant the ideas of dinosaurs and the Big Bang into our minds. Of course, Berkeley, who published his ideas in 1710 and 1713, knew nothing of these later theories.

Samuel Johnson made two criticisms of Berkeley's idealism. First, he kicked a boulder, hard enough to hurt himself, saying, "I refute him thus!" But, according to Berkeley's account, Johnson's pain is entirely predictable. His second, far more serious criticism is that the anatomy of human sensory organs explains our experiences, at least in part. Our hearing is muffled when our ears are blocked with wax, for instance. But sense organs are wholly irrelevant according to Berkeley's system, because God plants experiences directly into our minds. This is a striking and specific instance of Berkeley's lack of engagement with (2) **Science**. Johnson's worthless first objection is frequently mentioned, whereas his second, substantial concern is usually ignored. I suspect this is due to the quotability of the former.

Regarding criterion (1) **Adequacy**: Berkeley's idealism scores well in the domain of the mind-body problem, because it provided a coherent, rational explanation.

As it stands, the theory doesn't satisfy criterion (9) **Naturalism**; but one can readily strip Berkeley's God of all its religious associations. With this done, God is simply another basic entity in the universe, albeit an extremely powerful one; such a being is similar to Spinoza's "God or Nature" (see for example, Audi, 1999, pp. 870-74). It could be argued (though some will object) that this variant of Berkeley's idealism is naturalistic.

Evaluating standard physicalism

I next turn to evaluating standard (i.e., non-panpsychist) physicalism in terms of the criteria above. The discussion will be fairly condensed: for more complete detail, refer back to chapter 1, where several representative variants of physicalism were discussed. The fact that these variants exist is evidence of the difficulties that physicalism has with the mind-body problem.

All variants of physicalism unequivocally satisfy (9) **Naturalism**.

Qualia non-realism

Because of the severe problems of accounting for qualia within the framework of physicalism, some standard physicalists adopt the position of qualia non-realism, denying that qualitative experiences such as yellowness and pains are genuine. Qualia non-realists argue either that: ***(A)*** qualia do not exist; or ***(B)*** qualia do exist, but are illusions.

Denial that qualia exist fails criterion (3) **Accepting basic facts**. This is fatal for those whose specific domain of concern is the mind-body problem. The assertion that qualia do not exist is absurd: pain is an undeniable fact of human existence, of which we all have some first-hand experience; it is not some theoretical construct that can be lopped off with Occam's razor. The same applies to other experiential qualities. It is not legitimate to deny such basic facts on the sole grounds that they cannot be easily fitted into our current understanding.

Qualia illusionism categorically fails criterion (4) **Plausibility**. The notion that all experiences are somehow 'illusions' is **incoherent**: for what is an illusion if not an experience (albeit one that does not correspond to any physical reality outside the brain)?

(As discussed in chapter 1, ***(C)*** epiphenomenalism is not strictly speaking a form of physicalism, but it is closely related. According to epiphenomenalists, qualia do exist; but they have no physical effects. Because of this latter assumption, their supposed existence, which is useless, is absurd. This position also comprehensively fails criterion (4) **Plausibility**.)

Qualia realism

Because of the above failures, we will eliminate qualia non-realism from further consideration. From now on, *standard physicalism* will be taken to be shorthand for qualia-realist, non-panpsychist physicalism. Thus redefined, standard physicalism asserts that experience, qualia, and suchlike indeed exist, but only in tiny portions of the universe. As we saw in chapter 1, Standard physicalism comes in two closely-related variants: There are ***(D)*** *identity theories,* in which experiential states are **one and the same** (recall that, in philosophy, the term *identity* is always used in this very strong way). There are also ***(E)*** *emergence theories,* in which somewhere along a continuous path of evolution or development, a creature experiences the first glimmer of qualitative experience – perhaps a tingling sensation.

Standard physicalism accepts the reality of qualitative experiences such as yellowness and pains. It is therefore satisfactory in terms of criterion (3) **Accepting basic facts**. The theory is excellent on criterion (2) **Science**, because science is essentially unaltered – standard physicalism is a metaphysical add-on that can be ignored when undertaking science.

Standard physicalism completely fails on criterion (4) **Plausibility**. Within this metaphysical system, ontologically grounded in a microphysics that is wholly characterised in terms of spatiotemporal behaviour, the very existence of mind is utterly fantastical. David Chalmers, beginning from the basis that we are physical systems possessing minds, went on to wonder whether *philosophical zombies* might conceivably exist. According to his definition, these zombies are supposedly physically and behaviourally exactly identical to us, but they entirely lack minds (Chalmers, 1997, p. 94 onwards). Arguments about whether or not such zombies are possible are extremely controversial and theory-laden. However, if we start from a different, strictly physicalist viewpoint, which takes only the physical properties of humans as given, then we can readily conclude that it is **non-zombies** which are outright incomprehensible. (I'm afraid I don't recall who first framed the argument in this manner.)

In my view, standard physicalism also fails on plausibility for another reason: the problem of **(IV)** *free will*. It is extremely implausible that humans do not possess libertarian free will: that we aren't genuinely agents, who can sometimes consciously and freely choose which action to take. Consensus opinion weighs heavily against me on this point. Proponents of (mere) compatibilist free will admit that this is, at first sight, a highly implausible position. Despite this, arguments by followers of the consensus **presume** physicalism and **conclude** that we cannot have more than compatibilist free will: libertarian free will is ruled out. However, in the context of **evaluating** physicalism, such arguments beg the question. (Chapter 9 will be devoted to an examination of free will.)

We next consider criterion (5) **Minimal brute facts**. Standard physicalism has what I believe are fatal difficulties here. I will deal with identity and emergentist theories separately, but ultimately their faults are similar:

Standard physicalist identity theories claim that the particular patterns of neural processes now happening in my brain are identical to the headache that I am experiencing at the present moment. Here we have an intractable problem that has already been discussed, under the heading **(I)** *The explanatory gap*. There is no possible explanation that starts from the physical description of goings-on in my brain, and that terminates in the necessity of me having experiences, qualia, 'what it is like', and so on. As we saw in chapter 1, Papineau's argument, which begins by accepting that we have such experiences, and which terminates by providing convincing reasons why our brain states do not feature in our experiences – this argument proceeds in the wrong direction for a physicalist **explanation** of experience.

Regarding emergentism: In this theory mind first emerged in a particular species at some point in evolutionary history. Moreover, mind emerges in each individual conscious creature, at some point in its development from a fertilised egg. Emergentists hope that, because this initial qualitative experience is extremely primitive, it is unproblematic. This is not the case. First, **any** initial experience – even one as primitive as a tingling sensation – is an inexplicable novelty: the wholly intractable explanatory gap cannot be avoided. The earliest glimmering of experience will never be observable in terms of changes in a creature's behaviour; but this undetectability does not amount to an explanation of how this original experiential event came about. Second, emergentists are not excused from having to account for the full-blown technicolour phenomenology of humans. (Presumably they would have to fall back on identity theory here.)

Galen Strawson (2006a) and I, (Ells, 2011) argue – against Philip Clayton (2004) – that the **radical** emergence of qualia is in no way analogous to the **unexceptionable** emergence of novel spatiotemporal, behavioural properties, such as the emergence of liquidity as a property of a sufficient amount of water. Paraphrasing Strawson: if Y emerges from X, then it must be the case that all features of Y "trace intelligibly back to X (where 'intelligible' is a metaphysical rather than an epistemic notion)" (2006a, p. 18).

Given the above, in both variants of standard physicalism, mind is a miraculous brute fact.

On criterion (6) **Minimal promissory notes**, standard physicalism makes a huge promise about the future development of science. Throughout its history to the present day, science has come to grips with the world solely objectively – in terms of observation, measurement, and (within psychology) in reports of what subjects said. The outstanding successes of science are owing to this narrow empirical focus. There is no apparatus that can measure consciousness directly, and no prospect of one. Scientists would have to discover a 'consciousness meter' before they could study qualia such as pain **directly**, and it is obscure as to how this might happen, even in principle; see the talk by David Chalmers (1996b). Might pains be studied **indirectly** (as invisible particles such as electrons are studied) in terms of their manifest effects? Because the physical world is causally closed according to physicalists, this could only be done if pains were physical. But to claim that in the far future the methods of science might be expanded so as to include qualia is to make physicalism so vague as to be vacuous. This problem, already mentioned under the heading, **(II)** *Hempel's Dilemma*, applies to all forms of physicalism.

Standard physicalism makes a further promissory note: Neither the identity nor emergentist variants have yet managed to solve **(V)** the *boundary problem*. As yet, neither theory has been developed to the point where it has even the beginnings of a lucid, rational account, in physical terms, as to how one draws the boundary between the experiencing and non-experiencing portions of the universe.

On criterion (8) **Clarity versus vagueness**, standard physicalism is vague because it is unclear as to what is meant by 'physical' and by 'matter' – problems **(II)** and **(III)** above.

There is also vagueness as to the specifics of the identity: What phenomenal states are associated with what brain states? As remarked earlier, emergence theories also have to answer this question.

At first sight, criterion (7) **Elegance and simplicity** is fulfilled extremely well. On closer inspection, because of the extensive difficulties mentioned, standard physicalism fares poorly.

Consider criterion (1) **Adequacy**: From the discussion above, it is clear that, for many reasons, standard physicalism gives a wholly inadequate account of the mind-body problem.

Some people may be tempted to argue: Physicalism is true; so mind must eventually be explicable in terms of physics; so the mind-body problem will one day be within the scope of physicalism. This argument is not valid because, in assuming physicalism to be true, it begs the question.

Evaluating panpsychism

Recall from chapter 1, that panpsychism is a particular form of physicalist qualia realism, but one in which mind is present everywhere in the universe – including within the ultimates of physics (say electrons). These are mind-like in their intrinsic nature. Moreover, mental properties like qualia are on a par with physical properties, such as mass or charge. Panpsychism is not compatible with emergentism, so here I will take panpsychism to be a kind of physicalist identity theory.

As with standard physicalism, panpsychism is good in terms of criteria: (2) **Science**; and (3) **Accepting basic facts**.

As to criterion (4) **Plausibility**: A crucial advantage of panpsychism is that it significantly reduces problem **(I)** *The explanatory gap*. In this metaphysical system, it is the fundamental nature of all matter that it has mind-like characteristics. Panpsychists sometimes make use of a metaphor: 'from the inside' all matter has a mind-like character; 'from the outside' all matter is physical in character. In each instance, these two 'sides' are aspects of what is one **identical** thing: a single piece of matter.

Similarly, panpsychism has removed most of the sting from problem **(II)** *Hempel's dilemma*. It is more plausible to claim that mental and physical properties are identical, since we are now assuming that both are present throughout the universe. However, work remains to be done in order to give the specifics of this identity.

Is it implausible that fundamental particles (say electrons) have experiences? Not really. First, as we go down the scale of ever simpler biological systems, it becomes less and less obvious to us that they have experiences. There is no principled argument that allows us to draw a boundary between experiencing and non-experiencing entities in the biological domain. Moreover, there is no principled argument that can draw a boundary within the inorganic domain.

Second, no procedure exists that allows us to test directly for the existence of experience – not even in the case of humans. (Medical tests for 'levels of consciousness' do not directly detect *consciousness* as the word has been defined in this book. Instead, they assess behaviours believed to be associated with it: the ability to converse normally; the ability to respond to a question; the ability to respond to a pinch, and so on. Such tests therefore **presume** that consciousness is causally effective.)

Another advantage of panpsychism is that it resolves problem **(III)** *The nature of matter*: The intrinsic mind-like character of matter distinguishes it from a mathematical abstraction. In resolving this problem, it is of no importance that we cannot have access to the experiences of other entities – not even to those of other humans: the fact that the distinction exists is sufficient.

On problem **(IV)** *Free will*, panpsychism does not, of itself, make progress. Causation is fully described by the physical laws. Mental causation, agency, and free will only exist in virtue of the supposed identity between the 'inner' mind-like aspect of matter, and the 'outer' physical aspects. Progress can only be made with the free will problem when the details of this identity are made specific.

On criterion (5) **Minimal brute facts**: In panpsychism, the fact that entities have mental properties is a universal, fundamental truth – analogous to the fact that particles have mass – it is not a brute fact. In this respect, panpsychism fares better than standard physicalist identity theories.

On point (6) **Minimal promissory notes**: Panpsychism has the great benefit that it eliminates **(V)** the *boundary problem*. Indeed, this is the major motivation for adopting panpsychism.

An additional problem for panpsychism is the *combination problem*, mentioned in passing at the end of chapter 1. How do human beings, composed of myriads of individual atoms, each of which is a centre of experience and agency, come to have that unity of mind that we know we have? There are some hopes that this problem is solvable. Panpsychists have developed several preliminary ideas that at least indicate the beginnings of pathways towards solving this problem; see for example Gregg Rosenberg (2004, Part II).

Regarding criterion (8) **Clarity versus vagueness**: At present, nothing is known about the experiences of physical ultimates. All we can say is that, in general terms, the experiences of the simplest entities will be minimal. The fact that there is a continuous spectrum of entities, at all levels of complexity, might eventually give us traction on this problem. The present total lack of clarity need not be fatal, though this problem will remain for the foreseeable future.

Another point needing clarification goes by the name of the *aggregate problem*: Is a molecule a centre of experience in its own right; or is it just an aggregate of atoms, each with its own individual mind? Similarly, is a rock, of itself, a centre of experience; or is it no more than an aggregate of its constituent particles? Panpsychists have given different tentative answers to this problem.

Many variants of panpsychism are under active investigation, see Brüntrup and Jaskolla (2017, Part II). A couple of examples are mentioned here:

Cosmopsychism is a variant of panpsychism in which the

whole universe is mind-like, see Brüntrup and Jaskolla (2017, chapter 4). Instead of the combination problem, there is now the *separation problem,* which asks how this one great mind separates into finite, individual minds, such as ours. Spinoza's metaphysics comes into this general category: see for example (Audi, 1999, pp. 870-71).

Russellian monism is the term for a general class of theories in which mind and matter (or mind and brain) are two aspects of a single underlying substance (see Bertrand Russell, 1927, Parts II & III). These are currently enjoying a resurgence in popularity.

On criterion (7) **Elegance and simplicity**, panpsychism performs quite well.

Criterion (1) **Adequacy** has again been postponed to last. Panpsychism is far more credible than standard physicalism in giving an account of the mind-body problem. As explained above, of the five major problems of physicalism: **(III)** *The nature of matter* has been resolved completely. In addition, **(V)** the intractable *boundary problem* has been removed entirely. Great progress has been made with both **(I)** *The explanatory gap* and **(II)** *Hempel's dilemma,* with prospects for still more. No advances have been made in problem **(IV)** *Free will.* The new problems that panpsychism introduces – notably the *combination* and *aggregate problems* – seem to be tractable, having room for progress. To sum up, panpsychism gives the credible foundations of an adequate solution to the mind-body problem.

Comparison Table

Table 2.1 evaluates four metaphysical systems in terms of the nine criteria and the extensive discussions above. These systems are Berkeley's idealism, standard physicalism, panpsychism, and pan-idealism – this last will be detailed later. Theories of qualia non-realism are excluded because they deny basic facts or are incoherent.

Criterion	Berkeley's Idealism	Standard Physicalism	Panpsychism	Pan-idealism
(1) **Adequacy** (mind-body)	1	0.1	0.6	1
(2) **Science** (adequacy)	0	1	1	1
(3) **Accepting basic facts**	1	1	1	1
(4) **Plausibility**	0.5	0	0.6	1
(5) **Minimal brute facts**	0	0	1	0.9
(6) **Minimal promissory notes**	1	0.1	0.5	0.8
(7) **Elegance & simplicity**	0.8	0.2	1	1
(8) **Clarity versus vagueness**	0.5	0.2	0.2	1
(9) **Naturalism**	0.3	1	1	1

Table 2.1: Evaluating four metaphysical systems. My scores for panpsychism are invariably either greater than or equal to those of standard physicalism. Your scores for these criteria may well differ from mine.

Scoring how far each system meets these criteria

The extent to which a particular metaphysical system satisfies a particular criterion is not a Yes/No decision. Instead, this is evaluated by assigning a score between 0 and 1 (roughly: 0 = 'not at all', 1 = 'perfectly', 0.8 = 'quite well', and so on). This is shown by the numbers in the body of Table 2.1. Although there is room for subjectivity, agreement between evaluators should be reasonably consistent, because these scores are based on the extensive discussions above. I have chosen identity theories to represent standard physicalism because, in my opinion, they are the most convincing representatives within this category: they are 'the pick of the crop'.

It should be evident that a metaphysical system that satisfies every criterion to the same or a greater extent than an alternative system is to be preferred to the latter. By this reckoning, panpsychism is a definite improvement on standard physicalism.

Weighing the criteria

The nine criteria are not of equal importance. This could be taken into account in Table 2.1 by adding an extra column that assigns approximate, order-of-magnitude weights to each criterion. My own choice of weighting is: 10,000 for (1) **Adequacy** (mind-body), (2) **Science** and for (3) **Accepting basic facts**; 1,000 for (6) **Minimal promissory notes**, and (9) **Naturalism**; 100 for criteria (4), (5) and (8); and finally, just 10 for criterion (7). It would be a useful exercise to draw your own table, and assign your own scores and weights. Your figures will be different. I doubt, however, that even large changes will significantly alter the final conclusion when comparing these metaphysical systems.

Criterion (1) **Adequacy** (in explaining the mind-body problem) is weighted heavily because it is the main focus of this book.

Some criteria might be regarded as being **essential**, effectively having infinite weight. Many hold this to be so for (2) **Science**. One should be cautious here: Late in his life, William James developed a panpsychist theory that was inconsistent with the science of his era (1909). He did so because after many years of effort he had failed to find a scientific solution to the mind-body problem, which was his main subject of interest. He could not deny the existence of mind, so he reluctantly concluded that the science of his era must be radically wrong. His investigation of the mind-body problem could only get off the ground once he ignored Edwardian science. In the decades following his death, science changed radically, in ways that were conducive to his ideas. From James' example we can see that – only when all else fails – is it legitimate to sacrifice the objective of consistency with science to the principle of (3) **Accepting basic facts** (mind certainly exists). For this reason, (2) **Science** has been given great, but not infinite, weight.

Comparison

Comparing Berkeley's idealism with standard physicalism we find that standard physicalism comes out ahead; but the advantage is not as overwhelming as many people might have assumed initially. Berkeley's idealism falls down badly on engagement with science. This is counterbalanced by standard physicalism's almost complete failure to come to grips with the mind-body problem. Standard physicalism has been thoroughly investigated over many years, and it is extremely unlikely that this position will improve significantly in the future.

Panpsychism has already been compared with standard physicalism, in the final paragraph of the section **Evaluating panpsychism**. The conclusion was that panpsychism is a huge improvement. Moreover, it is still under active development, with great progress still being made.

A defence of metaphysics

Over the previous century, metaphysics (and philosophy in general) has come under severe criticism. Here I wish to list and respond to these critiques, which come under several headings.

- *Metaphysics is meaningless*

It is claimed that, because metaphysical systems are not empirical, and so cannot be tested, there is no rational basis for comparing them. But, as has been shown at length, well-founded comparisons can still be made, albeit using some personal judgement.

Our universe has an authentic character – in other words an objective metaphysics. There is a fact of the matter as to whether the cosmos is physics-like or mind-like in its fundamental nature. Just as physicists with their current theories attempt to approach the objective physics of the cosmos; so philosophers attempt in their theories to discover more about the world's objective metaphysics.

- *Philosophy makes no progress*

In contrast to the rapid advances in science, philosophy has made little progress; instead, it has re-ploughed the same territory again and again over the course of millennia. Against this I would argue that the almost absolute requirement for consistency and engagement with present-day science is a fairly novel criterion. Moreover, this weighty criterion will lead to rapid collaborative progress in both philosophy and in science.

- *Philosophy has been replaced by science*

Steven Hawking was an eminent scientist who held this view. In a talk he said, "Why are we here? Where do we come from? Traditionally, these are questions for philosophy, but philosophy is dead. Philosophers have not kept up with modern developments in science, particularly physics" (2011, at 1:00). I agree with part of Hawking's critique: philosophers have a responsibility to engage with present-day physics. Though not a professional philosopher, let alone a scientist, I have tried to do this throughout the course of the book.

Nonetheless, Hawking is wrong in his main contention that physics has entirely replaced philosophy. If you have read about quantum mechanics – even at a popular level – you will know that is has several interpretations, giving radically different pictures of what the universe is actually like, but which are identical in terms of experimental results to-date. Such interpretations are metaphysical systems, and to debate them is to engage in philosophy.

A scientist might respond that such debates are best left to scientists, and that philosophers have nothing useful to contribute. Moreover, all extant (respectable) interpretations of quantum mechanics are essentially physicalist. In his talk, Hawking implicitly touched upon his physicalism, saying, "... we humans, who are ourselves mere collections of fundamental particles of nature ..." (2011, at 23:30). This brings us to ...

- *Science has proven that physicalism is inevitable*

This would be the case if physicalism was the only metaphysical position consistent with current science, as a great many scientists, Hawking included, contend. Some philosophers agree:

> Physicalism is a thesis about the nature of the world that we have considerable and perhaps even overwhelming reason to believe. ... Those who deny physicalism are not making a conceptual mistake, but they are, nevertheless, flying in the face not merely of science but also of scientifically informed common sense.
> (Daniel Stoljar, 2010, p. 13)

To refute this contention, the purpose of this book is to develop a contrasting metaphysical system, *pan-idealism*. As its name suggests, it is a form of idealism rather than physicalism. But it is just as consistent with, and as fully engaged with current science, as any form of physicalism. Moreover, in terms of providing an adequate account of the mind-body problem, it is far superior to any existing form of physicalism – even panpsychism.

There is a danger in rejecting or ignoring philosophy: It is all but inevitable that everyone possesses a metaphysical position – and will act according to it. If we fail to investigate alternatives then we are liable to assume a poor position implicitly, and be reluctant to give it up. To be specific, I believe that physicalism is untenable and beyond any credible hope of correction: detailed arguments have been presented in this book (and also in Ells, 2011). Novel and under-explored metaphysical positions await our investigation.

Introducing pan-idealism

Pan-idealism is a metaphysical system which has the great advantage that all of the intuitions of the supposedly simplistic concepts of the 'watering a plant' example given at the start of chapter 1 can be shown to be essentially true. It is best introduced in the context of the four major facets of the mind-body problem, which were listed there:

1. *At first sight there are two, contradictory worlds: the physical and the mental*

1.1. *Which world are we living in?*

We are living in a universe that is, at its foundational level, entirely mentalistic in character. Hence pan-idealism is a form of idealism. The ultimate constituents of the universe, which I call *experients*, have the following characteristics: **(a)** They are (usually primitive) centres of experience; **(b)** They have percepts of other constituents; **(c)** A percept is intrinsically qualitative, and the qualia delineate structures within it; And **(d)** each experient is able, by an act of volition, to change its own percept, and also the percepts of others.

Despite this, pan-idealism is a realistic theory: The universe has an objective physics, to which current physical theories are an excellent approximation. In pan-idealism – in major contrast to panpsychism (or to physicalism in general) – objective physics is a secondary rather than a fundamental actuality. It is obtained, as we shall see in chapter 8, by combining the percepts of all experients in the most consistent manner possible.

In more detail: At any given moment, the percept of each experient has a certain mathematical structure (for example, Locke's visual percept of his desk had a particular mathematical structure). There exists a mathematical object that combines the structures of the percepts of all experients in the most consistent manner possible. This mathematical object – obtained from the percepts of all experients – is **identically** the physical structure of the desk (or, in a more general argument, of the world).

To make this clearer, it is somewhat analogous to a group of people determining the physical structure of a desk, by comparing their percepts of it. In pan-idealism there are a couple of differences: experients are omnipresent in the universe; and the desk itself is comprised of a multitude of experients.

With objective physics characterised thus, **(d)** amounts to saying that experients can, by an act of volition, change their physical state. For example, when I choose to pick up a pencil, very soon afterwards my percept changes to one in which I perceive myself to be holding a pencil. Witnesses will have corresponding percepts. Together, our revised percepts, mathematically combined, **constitute** the new physical situation.

The catalogues of real, individual entities are identical in pan-idealism and physicalism. The difference is that, according to pan-idealism, these entities are all experients. Pan-idealism is thus a non-physicalist cousin of panpsychism.

1.2. *What is the relationship between a person's mental state and their brain state?*

A person's brain consists of a hierarchical system of experients. His or her physical brain state is determined by (supervenes upon) this mentalistic hierarchical system. Supervenience proceeds in the opposite direction to the way it is claimed to go according to physicalism. In chapter 8 it will be shown that there is no explanatory gap in pan-idealism.

2. *How can minds perceive the physical world?*

2.1. *How do structural physical facts (a book is in front of you) become structural mental facts (your visual percept is of a book)?*

Experients have perception as a fundamental characteristic – without any need for sensory organs. The purpose of such physical organs (which are themselves composed of myriads of primitive experients) is to integrate information, so as to allow for more coherent and complex percepts.

2.2. *How can we explain qualia?*

Qualia are fundamental in pan-idealism.

3. *How can minds, as agents, cause changes in the physical world?*

3.1. *How are mental and physical causation related?*

How minds, as agents, cause physical changes was explained near the end of discussing 1.1. In pan-idealism, the **only** form of causation is volition. There is no physical causation: what appears to us as being physical causation is just a reflection of underlying, volitional causation. The mathematical regularities of the laws of objective physics reflect the stereotypical volitions of rudimentary experients, which are, in many of their properties, identical.

3.2. *Can we make free choices? If so, how?*

According to pan-idealism, there is no causation in objective physics; instead, all causation is volitional. These facts permit primitive experients to have wanton freedom, which is based on their percepts. Experients are fundamentally mind-like but, at the secondary level of objective physics, the foregoing can be witnessed in their empirically random behaviour. (By *empirically random*, I mean that, in certain situations, no experimental probe will enable us to predict with certainty how a particle will act.) Pan-idealism also allows human beings especially, and also higher animals to a limited extent, to possess authentic (what philosophers call libertarian) free will. This will be explained in detail in chapter 9.

4. *How did consciousness evolve in a universe that was initially lifeless for billions of years?*

As already mentioned, pan-idealism is closely related to panpsychism in that experients are abundant everywhere in the universe throughout the entire course of its history. They are even plentiful in the vacuum of space.

Pan-idealism and watering a plant

Pan-idealism, as its name suggests, is a form of monism. How then can we explain the apparent dualism of Figure 1.1?

The answer is that the physical world is real, but is wholly secondary. (Even the correct, and so-called 'fundamental' laws of objective physics have this character.) How this comes about is briefly sketched in the answer to question 1.1 above. Physical entities are, or are composed of, experients. The dualism of Figure 1.1 is thus explained as being only superficial: Closer analysis shows that the physical can be reduced in its entirety to the mental.

Evaluating pan-idealism

For now, the high scores given to pan-idealism in Table 2.1 will have to be taken on trust. I believe them to be justified even though the theory is far from complete. You will be able to make your own assessment after reading Part III. A novel and decisive advantage of pan-idealism is that it is thoroughly and precisely integrated with current physics.

As a preliminary to carrying out this project, the task of Part II is to review some essential concepts of contemporary physics.

Part II

Fundamental physics

Chapter 3

Classical physics

Before the development of quantum mechanics during the twentieth century, all theories of physics – including Newtonian mechanics, and the special and general theories of relativity – were classical in character. This chapter outlines these theories and describes the features they share, which motivates their common designation as *classical*.

Newtonian mechanics

In the ancient world, the most influential model of the heavens was a geocentric one: with the Earth at the centre of the universe, and with the Sun, Moon, and planets (then thought of as 'wandering stars') orbiting the Earth. Enclosing all this was the sphere of fixed stars, whose apparent nightly rotation was caused by the diurnal rotation of the Earth about its poles. The theory was finalized by the second-century mathematician and astronomer Claudius Ptolemy. His system was rightly influential as it correctly predicted the apparent positions and motions of celestial objects, including the planets, at least insofar as these had been measured during that era. A major disadvantage was that each planet moved in a complicated pattern of circles-upon-circles (called *epicycles*) – and also in accordance with other, similarly artificial, contrivances. These had to be concocted for each planet individually, in order to make the theory agree with observation.

Just before his death in 1543, Nicolaus Copernicus published his treatise, *On the Revolutions of the Celestial Spheres*, in which he argued that the planets orbit the Sun. In broad terms this gave a more natural account of the apparent motions of the planets. For example, the outer planets Mars, Jupiter and Saturn (Uranus and Neptune had not yet been discovered) sometimes went backwards (so-called *retrograde motion*), counter to their usual motion against the background of fixed stars. Copernicus had a natural explanation for this: Retrograde motion occurs then the Earth overtakes the outer planets as they follow their various orbits around the Sun. Copernicus' theory thus explained, at least in approximate terms, when, why, and to what extent retrograde motions occur. However, to obtain agreement with observation, Copernicus still needed to retain epicycles and other such contrivances in his theory; in fact, he required more epicycles than Ptolemy.

The astronomer Tycho Brahe (1546-1601), using his unaided eye, made comprehensive observations of the positions of the planets against the background of stars, by using large instruments of hitherto unprecedented accuracy. Johannes Kepler was a senior assistant of Brahe and he showed by a close mathematical analysis of the latter's observations that: **(a)** planets travelled in elliptical (almost circular) orbits around the Sun, with the Sun being at one focus of this orbit; **(b)** each planet travelled faster when closer to the Sun, in such a way that the line joining the planet to the sun swept out equal areas in equal times; and **(c)** that planets further from the sun had a longer orbital period 'year' (Kepler gave a specific mathematical rule, namely that the square of the orbital period of a planet is proportional to the cube of the semi-major axis of its orbit). He published these three laws as he discovered them, between 1609 and 1619.

Although Kepler's three laws are mathematically precise, he did not give any physical motivation for them – they just worked. When, in 1679, Robert Hooke wrote to Newton asking if such a physical motivation could be given, Newton was set to thinking, and this resulted in his publishing *Principia Mathematica* in 1687. Newton showed, among other things, that each piece of matter was attracted to each other piece by a force proportional to the product of their masses, and inversely proportional to their distance apart:

$$F = GMm / r^2$$

Here M and m are the two masses; r is the distance between them; and G is the constant of proportionality (known as *Newton's gravitational constant*).

Newton proved that this law (together with others) could be used to derive Kepler's three laws. There are other benefits. Some planets (particularly Jupiter and Saturn) are massive in their own right; for this reason, Newton's law can also be used to predict slight perturbations of planets from Kepler's elliptical orbits. Newton explained the tides in terms of the gravitational attraction of the Sun and the Moon acting upon the matter (specifically the water in the oceans) of the Earth.

Newton's laws also apply at small scales, and can be used to predict and explain the trajectories of balls and falling objects. For instance, Newton's laws make clear why (in situations where we can neglect air resistance) all objects fall with the same acceleration when dropped. Such facts were known from the experiments of Galileo, who published his results in 1638. It is interesting to note that, in 1971, Apollo 15 astronaut David Scott demonstrated dropping a hammer and a feather on the Moon – they landed together.

Newtonian space and time

In his *Principia*, Newton developed the first, fundamental theory of Nature. Anyone attempting such a task cannot begin from nothing: they must, of necessity, base their work on certain physical principles that cannot be proven. Newton did this explicitly, and formalised our intuitive conceptions of space and time:

> Absolute space, in its own nature, without relation to anything external, remains always similar and immovable... Absolute, true, and mathematical time, of itself, and from its own nature, flows equably without relation to anything external.
> (Newton, quoted in Misner, Thorne & Wheeler, 1973, p. 40)

So, Newton defined space and time carefully. Space exists in its own right as a physical container, independent of any of its contents. It has the flat, three-dimensional structure, familiar from Euclid. Time exists irrespective of whether or not any events occur within it.

Inertial reference frames

Newton's laws only work in their simplest form in what are called *inertial reference frames*. According to Newton, these are frames that are either at rest in, or in uniform motion with respect to, absolute space. In contrast, when working in non-inertial frames, extra terms called 'fictional forces' have to be added.

The simplest example of a non-inertial frame is a rotating one. Imagine a stationary bucket, half full of water. The surface of the water is approximately flat. Then the bucket is rotated on a turntable. The surface of the water is now concave, being lower at the centre, and higher at its edge. In the non-inertial reference frame that is fixed with respect to the turntable and bucket, there is a 'fictional' centrifugal (outward directed) force, which pushes the water up the sides of the bucket.

With respect to the ground, which is approximately an inertial reference frame, the rotation of the bucket induces a rotation of the water inside it. The circular motion of the water means that it must be undergoing a centripetal (inward directed) acceleration. The inertia of the water causes it to rise up the sides of the bucket.

There are thus two accounts of the behaviour of the water:

in the non-inertial frame there is a 'fictional' force. In the inertial frame there are accelerations with respect to that frame. Although the accounts differ, they predict the same results – in this case the shape of the water's surface. Newton's laws make the correct predictions in inertial frames, which are fundamental to the theory. All calculations can in principle be performed using inertial frames alone. But if a non-inertial frame is chosen for convenience, then the 'fictional' forces are contrived so as to give the same results. A reference frame fixed on the ground is not quite inertial because of the rotation of the Earth but, for the bucket example, the 'fictional' forces are small enough to be neglected. More refined experiments will reveal them, as will natural phenomena, such as global patterns of winds.

The forces you experience when riding on a rollercoaster give further examples of non-inertial 'fictional' forces, which arise because as seen from the ground the car you are sitting in is being wildly accelerated in different directions. The concept of inertial frames continues to be relevant in all of the classical theories to be discussed in this chapter.

Thermodynamics

Classical *thermodynamics* is a statistical theory that, among other things, explains the relationships between the temperature, pressure and volume of a gas by regarding it as a system of molecules moving at random. The statistics are averages that summarise our ignorance. In contrast, suppose we knew the sizes and the masses of all the molecules; and suppose also we knew their exact positions and velocities at some initial time. Under these circumstances we would – at least in principle – be able to use Newton's theory to compute future temperatures, pressures and volumes. So: **if** we were able to ignore the sheer impracticability of discovering all these facts, **then** the theory might be regarded as no more than a convenient shortcut. But of course, we cannot know these facts or perform these calculations, so classical thermodynamics remains an essential tool.

Electrodynamics

Classical electrodynamics is the theory that unified electricity and magnetism as coming under one set of laws. It was developed descriptively by Michael Faraday during the first half of the nineteenth century. He proposed that there exist electrical and magnetic fields of force that extend throughout space, and his view was eventually accepted. James Clerk Maxwell used differential equations to describe the theory mathematically in his book *A Treatise on Electricity and Magnetism,* published in 1873. He described magnetic and electrical fields as being physically coupled: one can induce an electric current to flow in a wire by waving a magnet nearby; and, conversely, an electric current flowing through a wire gives rise to a magnetic field that surrounds it. Light was eventually understood to be an electromagnetic wave (undulating disturbance) in these coupled fields. Visible light consists of those wavelengths we happen to be able to see. Light also exists at shorter wavelengths, for example X-rays; and at longer wavelengths, such as microwaves and radio waves.

Maxwell's laws imply that the speed of light in a vacuum (called c) is a constant. Its exact value is c = 299,792,458 metres per second. An excellent approximation is given by c = 300,000 km per second.

Special relativity

Special relativity arose from the failure of attempts to reconcile Newtonian mechanics with classical electrodynamics. According to the latter theory, the speed of light is a constant c. The most natural assumption is that this constant velocity c is to be measured with respect to Newton's absolute space. If this is the case, we should be able to discover experimentally how fast we are moving with respect to absolute space.

In principle, the experiment is as follows. Have a light source that is in a fixed position in our lab. Make the source flash for a moment. Consider that the lab and all its contents are moving with velocity v in direction d with respect to absolute space. For simplicity assume that v is smaller than c. By the addition rule for velocities (which is part of Newton's theory), the velocity of light in direction, d, should be c-v, as measured in the lab. In contrast, the measured velocity of light in the opposite direction, -d, should be c+v. If we measure the speed of light in many possible directions, we expect to find a range of values. The difference between the maximum and minimum values, (c+v) - (c-v) = 2v, should give us the velocity v of the lab with respect to absolute space. Moreover, the direction in which the speed of light is at a minimum should be the direction, d, in which the lab is travelling with respect to absolute space. This experiment gives a null result, as Maxwell's laws predicts. But this failure might simply mean that v << c: The conclusion, that we are moving very slowly with respect to absolute space, would seem quite plausible, given what was known during that era.

In 1887 Michelson and Morley performed a much more sensitive experiment, using apparatus named after them. The experiment relied on the fact that light has wave-like characteristics. It was then believed that these were waves in a substance, the *luminiferous aether*. This substance was supposedly extremely rigid as regards to its interaction with light; in contrast, it did not interact at all with any other matter – it could be regarded as being infinitely tenuous. As the Earth orbited the Sun over the course of several months, it should have been possible to obtain results that varied by a measurable amount, according to how light within the Michelson-Morley apparatus crossed the aether. The details of this apparatus and experiment are unimportant here – they relied on similar principles and assumptions to those required for the thought experiment of the previous paragraph.

When Michelson and Morley's results came in, neither absolute space, nor motion through the aether, were detected. The conclusion was that Newtonian mechanics and classical electrodynamics could not be reconciled – at least, not without some essential amendments.

Ad-hoc solutions

The initial, ad-hoc amendment was minuscule, albeit extremely strange. The null result could be explained if it was supposed that the Michelson-Morley apparatus physically shrank in the direction of its travel through the luminiferous aether. The amount of shrinkage was a function of the speed of this travel, but the same shrinkage occurred irrespective of the material out of which the apparatus was made.

The *Lorentz Transformation* is as follows: Imagine person A is at rest at the origin of an (inertial) coordinate system (x, y, z, t). Person B is travelling with velocity v in direction x, and is at rest at the origin of her (inertial) coordinate system (x′, y′, z′, t′). They pass one another when both of their clocks read zero: $t = t' = 0$.

Let $\gamma = 1 / \sqrt{(1 - v^2/c^2)}$. This γ (gamma), called the *Lorentz factor*, has the following properties: when $v \ll c$, $\gamma \simeq 1$; when $v = c/2$, $\gamma \simeq 1.15$; when $v \simeq c$, $\gamma \simeq \infty$. To put this into words: The Lorentz factor γ remains close to 1 until v becomes comparable with the speed of light. Even when v is half the speed of light, γ is not much bigger than one. As v approaches c, γ tends to infinity. The Lorentz transformation can now be written:

$$t' = \gamma(t - vx/c^2)$$

$$x' = \gamma(x - vt)$$

$$y' = y \qquad z' = z$$

As Hendrik Lorentz initially envisaged this transformation in 1895, person A was at rest with respect to the luminiferous aether, with (x, y, z, t) being 'true' coordinates; whereas (x′, y′, z′, t′) was the 'physically distorted' coordinate system of person B, who was moving with respect to this aether. Lorentz was able to show that his transformation explained (or at least explained away) the null results of the Michelson-Morley and similar experiments.

The inverse Lorentz transformation is:

$$t = \gamma(t' + vx'/c^2)$$

$$x = \gamma(x' + vt')$$

$$y = y' \quad z = z'$$

As you can check, we obtain the inverse Lorentz transformation from the original formulae by replacing –v with +v throughout; and also by removing primes on the LHS, and adding them to the coordinates on the RHS. There is a specific symmetry here: From the viewpoint of person B, person A is travelling with velocity –v (in the direction minus x′).

Einstein's concept of Special Relativity

The supposed properties of the luminiferous aether are extremely peculiar – almost infinite rigidity combined with infinite tenuousness. The Lorentz contraction is also bizarre when regarded as a physical compression of the apparatus: first, the compression is independent of the material content of the apparatus; second, it conspires to conceal the motion of the apparatus through the aether. It is not at all clear as to whether or not it would be possible for other apparatus, constructed on different principles, to detect such motion. Moreover, the luminiferous aether was experimentally undetectable.

Albert Einstein, influenced by Ernst Mach, wanted to eliminate from physics putative entities that could not be studied experimentally. He therefore strove to put physics on new foundations that rejected the notions of absolute space and time; and also of the luminiferous aether. He proposed that two things were fundamental: First, the laws of physics are identical in all inertial reference frames. Second, the speed of light in a vacuum is a constant, c, in agreement with classical electrodynamics.

Einstein did not want to evoke concepts such as 'time' or 'distance' without specifying concrete physical devices by which these could be measured. At this preliminary stage of developing a fundamental theory, such measuring devices must be the simplest possible that could do the job – at least in principle. A conventional clock, for example, could not even be described yet – it is far too complicated. In the preface to his "popular exposition," Einstein states that he presents his ideas "in the sequence and connection in which they actually originated" (1920, p. v).

Distance definition: How can we measure the distance between two locations that are stationary in our reference frame? Einstein suggests that the experimenter is equipped with a rigid rod, and the number of rod-lengths of a straight line between the locations is used to measure the distance (1920, pp. 1-5).

Simultaneity: Suppose that lightning strikes at two distant locations. How is an observer to judge whether or not these strikes are simultaneous? Einstein proposed the following somewhat artificial situation: suppose the observer happens to be located at the position exactly midway between the locations (as defined by the distance definition); suppose further that the observer can view both locations continuously (by having a pair of suitably oriented mirrors directly in front of him). If the observer sees the flashes occur simultaneously in his mirrors, then the flash events indeed occurred *simultaneously* in his frame of reference (Einstein, 1920, pp. 21-4). The motivation for this definition is that the speed of light is a constant c in any reference frame. In particular, it can be inferred that both of the flashes occurred t = d/c seconds before the observer saw them, where d is the distance of each location from the observer.

From his two fundamental principles, Einstein was able to **derive** the Lorentz transformation (1920, Appendix I). The details will not be given here, but his famous thought experiment in which a train travels at uniform speed v along an embankment (1920, pp. 25-7) is a relevant partial solution; see Figure 3.1.

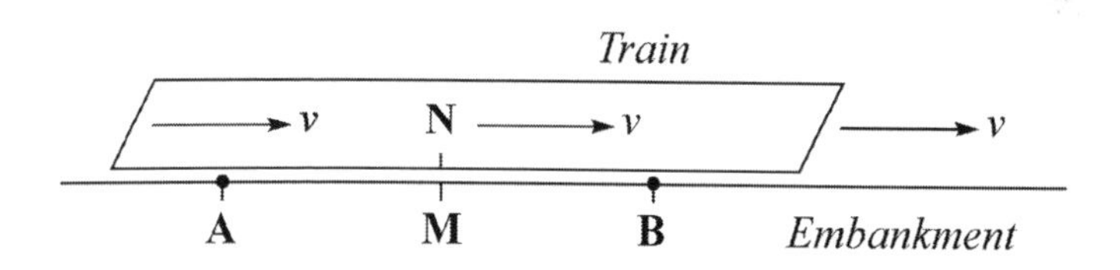

Figure 3.1: Einstein's train, showing that simultaneity is only meaningful when expressed relative to a given coordinate system (1920, p. 25). Lightning flashes occur at A and B, which are simultaneous as seen by person M. What does person N see?

Person M stands on the embankment midway between locations A and B. Using the mirror device, he sees lightning strike simultaneously at these places. Person N is on the train, and passes M, at the moment when (according to M's companions, standing on the embankment at locations A and B) the lightning strikes. Companions A and B each witness the lightning strike at their location immediately – a little before M sees it.

Suppose N leaps from the train to join M as they pass one another. N will be standing next to M, and both will witness the lightning striking simultaneously.

But suppose instead that N remains on the train. As Einstein explains:

> He is hastening towards the beam of light coming from B, whilst he is riding on ahead of the beam coming from A. Hence the observer [N] will see the beam of light emitted from B earlier than he will see that emitted from A. Observers who take the railway train as their reference body must therefore come to the conclusion that the lightning flash B took place earlier than the lightning flash A. We thus arrive at an important result:
> [...] Every reference-body (coordinate system) has its own particular time; unless we are told the reference-body to which the statement of time refers, there is no meaning in a statement of the time of an event.
> (Einstein, 1920, p. 26)

Contrast this with the absolute time of Newton's theory. As already mentioned, a more detailed and quantitative analysis enabled Einstein to derive the Lorentz transformation from his two principles.

The Lorentz transformation is now to be understood as formulating how person A, who was fixed with respect to a given inertial coordinate system (x, y, z, t), could relate his measurements to those of person B, with her corresponding inertial coordinate system (x′, y′, z′, t′). In contrast to Lorentz, Einstein was the first to understand the transformation as being one between a pair of different **but equally valid** coordinate systems, A and B: hence the term "relativity". The symmetry between the Lorentz transformation and its inverse was a consequence of Einstein's first principle.

Key facts

A *rest frame* of an object is any inertial coordinate system with respect to which the object is at rest. Given this definition, here are some key facts about special relativity that will be explored more fully later:

Lorentz contraction: Suppose an object, say a railway carriage, has a certain length $\Delta x'$, as measured in its own rest frame R′.

Let Δx be the length of the carriage as measured from the embankment. In order for this measurement to be valid, the positions of the ends of the carriage must be measured simultaneously (in the rest frame R of the embankment): i.e., $\Delta t = 0$. As seen from the embankment the carriage is moving with velocity v in direction x. From the Lorentz transformation it can be shown that:

$$\Delta x = \Delta x' / \gamma \quad \text{where } \Delta t = 0$$

So, the length of the moving carriage **as measured from the embankment** (Δx) is less than the rest-length of the carriage ($\Delta x'$) by a factor $1/\gamma$. This is the *Lorentz contraction*. For example, when $v = c/2$, $\gamma \simeq 1.15$; so the carriage (and likewise the entire train) has contracted to $1/1.15 \simeq 0.87$ of its rest-length, as measured from the embankment.

The occupants of the train will perceive no distortion in the dimensions of their carriage. But if they look out of the window, they will measure the embankment as being contracted by the same factor $1/\gamma$ as it hurtles in the direction $-x'$.

The formula for the Lorentz contraction applies to any rigid object moving with velocity v in direction x.

Time dilation: Suppose a clock is in a fixed position within the moving carriage ($\Delta x' = 0$), and the interval between its ticks is $\Delta t'$, as measured by a passenger in the carriage. What is the interval (Δt) between successive ticks, as this is measured from the embankment? From the Lorentz transformation it can be shown that:

$$\Delta t = \gamma \, \Delta t' \qquad \text{where } \Delta x' = 0$$

In other words, as measured from the embankment, the moving clock has slowed down by a factor of γ. This effect is called *time dilation*. For example, if v = c/2, then the time dilation is $\gamma \simeq 1.15$.

Time dilation affects more than just timepieces; **all** physical processes will have slowed down as measured from the embankment. The passengers on the train move in slow motion; their hearts beat more slowly; they even age more slowly; objects that are dropped fall to the floor of the carriage more slowly – all as measured from the embankment.

None of the train passengers would notice anything unusual about the watches of their fellow travellers, nor about rate of any physical process on the train. But if these passengers measured the watches of the people on the embankment, they would find them running slow. The passengers would also see the weird slowing of all physical processes on the embankment – as measured from their rest frame of the train.

Are length contraction and time dilation merely illusions? No – they have real physical effects. For example, muons are particles that have short lifetimes after being created in a laboratory. Very fast-moving muons are created when cosmic rays hit the Earth's atmosphere. These latter muons have a much longer lifetime, as measured in the laboratory frame.

A measurement is only meaningful if the inertial reference frame in which it is made is reported. Provided this is done, all such measurements are equally valid.

Addition of velocities: Suppose person N on the train bowls a ball with velocity w, as measured by him. The direction of the ball is along the corridor, towards the engine. What is the velocity u of the ball, as measured from the embankment? If Newtonian physics were true, this would just be:

$$u = v + w$$

In special relativity, it can be shown that the answer according to the Lorentz transformation is given by:

$$u = \frac{v + w}{1 + (vw/c^2)}$$

This is clearly a smaller number than before. In special relativity it turns out that nothing can travel faster than light. This being so, if both v and w are both $< c$, then $u < c$.

Minkowski space

The Lorentz transformation has been given here for uniform motion with velocity v in the x-direction. But we could derive similar formulae for different velocities w in **any** direction. In each case, Lorentz contraction will be different both in size and direction. And time dilation will be different. With all these variations, at first sight it seems difficult to conceive of anything that remains constant, and upon which we can build our intuitions. Mathematician Hermann Minkowski developed a geometric concept, the interval, that solved this problem.

Geometric prelude: In ordinary three-dimensional space we have the intuitive concept the *distance* d between any pair of points A and B. You will be able to check informally, by putting the rules into English, that d has the following properties, for arbitrary points A, B, C:

$d(A, B) = d(B, A)$

$d(A, A) = 0$, but $d(A, B) > 0$ when B is different from A

$d(A, C) \leq d(A, B) + d(B, C)$
This is called the triangle inequality

(In fact, these rules apply to Euclidian spaces of any number of dimensions.)

For many centuries geometry was studied using ruler and compasses – effectively using the concept of distance just given. Then, in 1637, Descartes published the idea of *Cartesian coordinates* (x, y, z). I am assuming that you are familiar with these coordinates, so the following is merely a refresher. The general idea is to set up an arbitrary point O as an origin, and to choose three directions x, y, z from this origin, at right angles to one another. One then specifies the coordinates of a point, say A, by stating how far one would have to travel from O, in each of these directions, in order to reach A. For example,

$$C = (1.7, 0, -3)$$

This means that to get from O to C we have to go distance 1.7 in the x direction, no distance in the y direction, and distance 3 in the direction opposite to z. With this rule, clearly O = (0, 0, 0). Now suppose we take arbitrary points A, B with coordinates A = (x_a, y_a, z_a) and B = (x_b, y_b, z_b). We can calculate the distance d(A, B) between A and B by taking the square root of the following formula:

$$d^2(A, B) = (x_a-x_b)^2 + (y_a-y_b)^2 + (z_a-z_b)^2$$

This is just Pythagoras' Theorem in three dimensions. For brevity, write (x_a-x_b) as Δx, etc. The above formula now becomes:

$$d^2(A, B) = (\Delta x)^2 + (\Delta y)^2 + (\Delta z)^2$$

A crucial fact is that **we arrive at the same answer** for the distance between points A and B **no matter what coordinate system we choose**. Recall that we can choose the origin arbitrarily. We can also choose the directions x, y, z arbitrarily, subject only to the constraint that these directions must be at right angles to one another. We say that distance is an *invariant*, because it remains constant irrespective of the Cartesian coordinates chosen. This argument is readily amended for spaces of any number of dimensions.

Minkowski space: We are now ready to discuss Minkowski's idea. Consider the universe as being a geometric object, called *spacetime*, having three spatial dimensions, and one time dimension. We want to visualise this so that the time coordinate is somehow the 'same size' as the space coordinate. We do this by using Maxwell's discovery that the speed of light in a vacuum is a constant, c. Instead of using t as the time coordinate, we picture spacetime using ct instead. As we shall see in a moment, this has the effect of making the slope of every light ray in spacetime equal to one when we make a sketch.

Choose a specific *reference frame* R (say 'the embankment') in spacetime. Frame R, with a coordinate system (x, y, z, ct) that is somewhat analogous to a Cartesian coordinate system. Choose a pair of *events* A B (say 'lightning flashes') that are characterised by being accurately located in both space and time. Events are physical happenings, but they are otherwise analogous to geometric points.

Events A and B have specific spacetime coordinates **relative to R**. Say $A = (x_a, y_a, z_a, ct_a)$ and $B = (x_b, y_b, z_b, ct_b)$. Minkowski's wonderful idea was to define the *spacetime interval* I(A, B) – or just *interval* for short – between A and B using the formula:

$$I(A, B) = (x_a-x_b)^2 + (y_a-y_b)^2 + (z_a-z_b)^2 - c^2(t_a-t_b)^2$$

Note that the square of the speed of light appears in the final term. Moreover, this term is negative. In contrast to the Euclidean squared-distance, the interval I can take both positive and negative values. Another distinction is that the interval can be zero even when $A \neq B$. Using the same abbreviations as before, we can rewrite the above equation as:

$$I(A, B) = (\Delta x)^2 + (\Delta y)^2 + (\Delta z)^2 - c^2(\Delta t)^2$$

Different people use different conventions. Some call the square root of I(A, B) the interval. This alternative convention has the disadvantage that the interval can then be a complex number. Moreover, it complicates the discussion of light cones to be given below. Using the convention here, the spacetime interval is somewhat analogous to distance squared. Another alternate convention concerns signs. Some people write –, –, –, + where +, +, +, – is written in the above equations. The convention given here, in which the space terms are positive, is called the *spacelike convention*. The alternative convention is called the *timelike convention*.

Minkowski's concept of the spacetime interval I is brilliant because it is **invariant** under arbitrary changes of inertial reference frame. Suppose I(A, B) = 12.34 in frame R. Choose a different inertial reference frame R′. The coordinates of A and B will be different in this new frame. Expressing A and B in R′ coordinates gives $A = (x'_a, y'_a, z'_a, ct'_a)$ and $B = (x'_b, y'_b, z'_b, ct'_b)$: Typically, $x'_a \neq x_a$, and so on for all the other corresponding coordinates of A and B. The same holds for the coordinate differences in the different frames: $\Delta x' \neq \Delta x$.

Despite this, if we plug these different (primed) coordinates into the expression for the interval, we still arrive at the same result! In our example, I(A, B) still equals 12.34 when calculated using the coordinates of the R′ inertial reference frame. The same applies to any inertial frame.

The geometry of spacetime

The interval gives the causal structure of special relativity. Suppose, as before, that there is a pair of events A and B having coordinates $A = (x_a, y_a, z_a, ct_a)$ and $B = (x_b, y_b, z_b, ct_b)$ with respect to inertial reference frame R. But now, in addition, suppose that I(A, B) = 0. Then:

$$0 = (x_a - x_b)^2 + (y_a - y_b)^2 + (z_a - z_b)^2 - c^2(t_a - t_b)^2$$

$$c \times \Delta t = \sqrt{((\Delta x)^2 + (\Delta y)^2 + (\Delta z)^2)}$$

The RHS is just the Euclidian spatial distance between A and B (ignoring time) in the given frame of reference R. So:

$$c = d(A, B) / \Delta t$$

In words: Within the coordinate system of inertial frame R, the spatial distance between events A and B, divided by the time interval between these events, is equal to the speed of light. This means that a flash of light emitted from the earlier of these events will reach the later event exactly. The earlier event could be the emission of a photon (particle of light) from an atom; and the later event could be the absorption of this photon by another atom.

Consider Figure 3.2. Distinct events A, B in spacetime such that I(A, B) = 0 are said to be *lightlike separated*. This concept is independent of any inertial reference frame because the interval is an invariant. Suppose we select a random event A in spacetime, and that a flash of light is emitted from it. We can imagine all of the future events that are lightlike separated from A. If we sketch them, we get what is called the *future light cone* of A. Similarly, if we imagine all of the past events that are lightlike separated from A, we can sketch the *past light cone* of A. Temporal ordering is unambiguous for lightlike separated events: In the Figure, observers in all reference frames will agree that event Lf occurs on the future light cone of A; and similarly, that Lp occurs before event A.

In the real world, a momentary flash of light will expand from the source as a sphere of ever-increasing diameter. This four-dimensional event cannot be pictured easily, so we discard the z-dimension (which is similar to x and y anyway). The sphere is reduced to a circle in any time slice. As the circle expands it produces a cone: the future light cone. The past light cone is the collection of all events such that light emitted from them can arrive at the given event A.

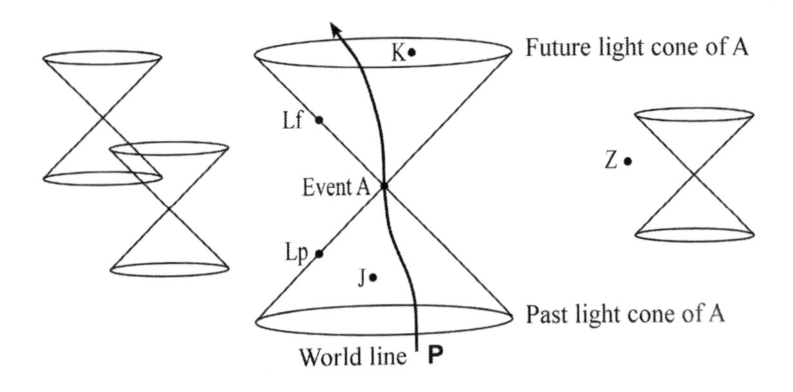

Figure 3.2: Light cones and the causal structure of Minkowski spacetime. This causal structure is invariant (independent of any inertial reference frame).

We still speak of 'cones' even when studying all four dimensions of spacetime: a light cone is a 4-D analogue of a cone. In principle, cones extend indefinitely into the past and future, but we are forced to truncate them in sketches. Likewise, for obvious reasons, only a selection of light cones can be drawn.

Figure 3.2 shows **P**: the *world line,* or path through spacetime, of some physical entity. This might be a person, a parrot, a planet, a pencil, or a particle (except a photon). The entity is regarded as essentially a point travelling through time – all other structure is ignored. For example, if it is a person, they are born at the root of the arrow and die at the tip. In special relativity, because nothing can travel faster than light, world lines must stay with the light cone at every point along its path. To continue the example, suppose event A is Peter's 30th birthday. Henceforth, Peter is destined to remain within the future light cone of A. He could, for example, reach event K but, as it turns out, he does not. But a guest at his party might reach event K.

The interior of the future light cone of event A is called the *absolute future* (of event A). K is a representative event within the absolute future of A. The upper portion of world line **P** also lies in this absolute future.

Similarly, the interior of the past light cone of event A is called the *absolute past* (of event A). J is a representative event within the absolute past of A. The lower portion of world line **P** also lies in this absolute past.

Events such that the interval between them is **negative** are said to be *timelike separated*. In the Figure, events A and K are timelike separated: I(A, K) < 0; likewise events A and J are timelike separated: I(A, J) < 0. Any events P_1, P_2 on any world line **P** are timelike separated. For example, the birth and death of Peter are timelike separated. Temporal ordering is unambiguous for timelike separated events: In the Figure, observers in all reference frames will agree that event K occurs later than A, and that J occurs before event A.

As an exercise, compare the concepts of lightlike and timelike separation.

Finally, if the interval between a pair of events is positive, they are said to be *spacelike separated*. Geometrically this means that each is outside the light cone of the other. We call the region outside the light cones of event A the *absolute elsewhere* (of A). In the Figure, events A and Z are spacelike separated: I(A, Z) > 0. Be warned that there is no well-defined temporal ordering for spacelike separated events: event A occurs before Z in some reference frames; afterwards in others; and they occur simultaneously in yet others.

Given any event A, we can partition spacetime neatly into six parts: (1) the event A itself; (2) its future light cone; (3) its absolute future; (4) its past light cone; (5) its absolute past; and

(6) its absolute elsewhere. Given A, another arbitrary event B belongs in exactly one of these regions.

In diagrams, light cones slope at 45°, as shown. This amounts to scaling figures such that the speed of light equals 1. For instance, if a certain distance on the vertical axis represents a year, the same distance on the horizontal axis would represent a lightyear. Moreover, with this convention, world lines always slope less than 45° from the vertical.

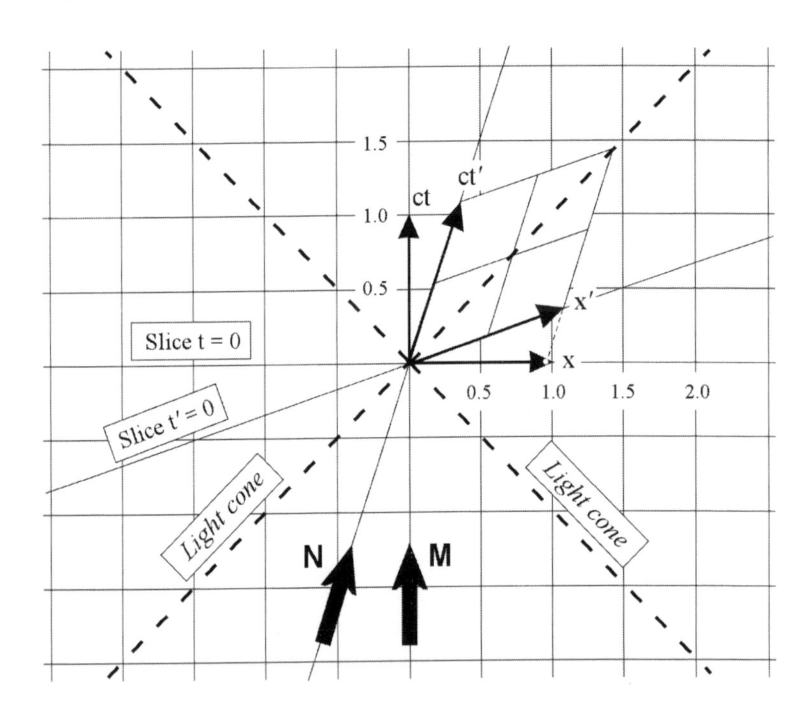

Figure 3.3: Inertial reference frames in Minkowski spacetime. Frame R is defined by coordinates (x, y, z, ct), but y & z have been omitted. Person M, with world line **M***, is at rest in this frame. Person N is at rest in frame R′ defined by coordinates (x′, y′, z′, ct′).*

Coordinates in Minkowski space

In Figure 3.2, everything was described using a coordinate-free approach. In Figure 3.3 we discuss the train example in terms of the inertial reference frames R and R′ of the embankment and train respectively. To simplify the drawing, dimensions y and z have been omitted. The origin O of both these coordinate systems is the event (at the centre of the sketch) where M and N pass one another.

The dashed diagonal lines are a cross section of the light cone at O. You can complete the cone in your imagination by rotating these lines about the vertical ct-axis.

The embankment frame: The inertial reference frame R, with the coordinates (x, y, z, ct) is the coordinate system fixed on the embankment, with origin at O. These coordinates are denoted in part by the (x, ct) grid that fills the Figure. Person M is stationary on the embankment, and his world line **M** is the line consisting of all points with R-coordinates (0, 0, 0, ct) for any t. In other words, **M** is the vertical line pointed to by the large arrow.

We can imagine person M having companions spread out along the embankment – their world lines might be the vertical lines of the grid. Suppose they synchronise their watches at the horizontal line marking the bottom of the Figure. Their clocks will remain synchronised from then on: they will agree as their world lines intersect every row of the grid. In particular, they will all agree that their watches read zero when they pass the gridline marked "Slice t = 0". In the diagram, this "Slice" is a horizontal line, but we can imagine it as a plane if we consider other observers standing in the fields in the ±y direction, into or out of the page. The "Slice" becomes a 3-D solid if we also consider the ±z direction. It consists of all events with R-coordinates (x, y, z, 0).

More generally, in any given coordinate system, we can fill the entirety of space with observers – stationary with respect to one another – whose watches are synchronised. The watches of these co-moving observers will remain synchronised indefinitely. (As we shall see later, in general relativity this rule can break down but, except in extreme situations, it remains fairly accurate for long time periods.)

The train frame: The reference frame R′ of the train is given by the apparently 'skewed' coordinates x′ and ct′. The other coordinates, y′ and z′, are identical to their unprimed counterparts, y and z. The unit square of the R′ coordinate system is shown as a skewed diamond. This is subdivided into four because each grid is subdivided into half units.

What does the speed v of the train happen to be in Figure 3.3? Consider the tower of three grid-squares above and to the right of the origin. The world line of N exactly crosses this tower diagonally. This means that, in 1.5 units of time, the train has travelled distance 0.5. So, in 1 unit of time, the train has travelled distance 0.5/1.5 = 1/3. But in one unit of time, light can travel a unit distance. It follows that v = c/3.

The angled line **N**, indicated by the large arrow, is the world line of person N standing on the train. **N** is the set of events with R′ coordinates (x′, y′, z′, ct′) = (0, 0, 0, ct′). According to person N, he is always standing at the spatial origin of his reference frame; he does of course experience the passage of time.

Person N can synchronise watches with his companions, co-moving on the train; their watches will remain synchronised. The skew line labelled "Slice t′ = 0" marks a moment of simultaneity agreed by all these passengers. The essential reason it is skewed has already been given in the Einstein's train argument.

The primed coordinates are distorted: This is only apparent: The coordinates x′ and ct′ appear to be squashed together in the Figure 3.3. In physical reality they are orthogonal to one another, just like x and ct. Moreover, as they are depicted in the Figure, arrows x′ and t′ are both drawn longer than arrows x and t, and by the same amount. (For v = c/3, this lengthening happens to be a factor of about 1.12, as you may check for yourself, using a ruler.) However, in reference frame R′, observers on the train would measure arrow x′ to be of unit length, and arrow ct′ to be of unit time.

More objectively, the spacetime interval between the spacetime events marking the tips and tails of each of the six spacelike arrows x, y, z, x′, y′, z′ is +1: this is true in all inertial reference frames. For example, arrow x is characterised by the origin O (the event at the tail) and event x that marks its tip: so I(O, x) = 1. For the two timelike arrows ct and ct′ we have I(O, ct) = -1 and I(O, ct′) = -1 respectively.

Time dilation and length contraction: When v = c/3, γ = 1.06, as you may check. This is still quite small. **Time dilation** may be seen almost immediately from Figure 3.3. A unit of time in N′s reference frame is marked by the ct′ arrow. This occurs later than (higher in the diagram than) the horizontal gridline marking ct = 1.

The **Lorentz contraction** is more subtle. Let us suppose that, when at rest on the embankment, the train has unit length, and coincides with the arrow x, as agreed by observers both on the embankment and on the train. The train backs up a long way, and then hurtles forwards, past observer M at speed v = c/3. At the instant t = 0 when the back of the train passes M, the entire train might seem to correspond to the spacetime arrow x′. But, in the diagram, arrow x′ is longer than arrow x. Why then don′t we have a Lorentz **expansion**?

The answer is that, in order to obtain a valid measurement of the length of the train, observers on the embankment must measure the positions of front and back ends of the train **simultaneously** in their frame. Measuring the position of the front of the train at the tip of arrow x′ makes this measurement too late. We must instead measure where the front of the train was at time t = 0. In Figure 3.3, there is a short, dotted line tracing the world line of the front of the train back in time to meet the x-axis.

Within the tip of arrow x, you can just see a little white dot. This marks where the front of the train was at time $t = 0$ according to observers on the embankment. According to them, the rear of the train passed the origin at this instant. The dot, being at a point where $x < 1$, indicates the Lorentz contraction.

Different viewpoints: It would be possible to redraw Figure 3.3 from the viewpoint of observers on the train. The light cones would be identical. Axis ct' would be vertical, and x' horizontal. The world line of N would be vertical. "Slice $t' = 0$" would be horizontal.

The world line of M would have $v = c/3$, but in the minus x' direction. Axes ct and x would be skewed, and would appear on the left side of the figure, as would the shaded grid patch between them. "Slice $t = 0$" would now be slanted, but the slope would be downwards, as one goes from left to right across the Figure. Drawn from the rest frame of the train, it is now the unprimed coordinates that are distorted.

All viewpoints are equally valid. Choose any inertial reference frame. Let this be the reference frame of an observer, say N. We can plot the spacetime diagram with N's world line vertically up, and their planes of simultaneity horizontal. Other coordinate frames will appear distorted, but this is just an accident of our choosing. Minkowski spacetime is perfectly symmetrical for all observers: the concepts 'vertical' and 'horizontal' do not exist for it. We are forced to use these concepts when we attempt to sketch Minkowski spacetime on paper.

Slices of simultaneity

I tried to explain slices of simultaneity in Figure 3.3, but I had to reduce these to a mere line, so the explanation might have been obscure. In Figure 3.4, I've eliminated the coordinate systems, and have reinstated the y-coordinate.

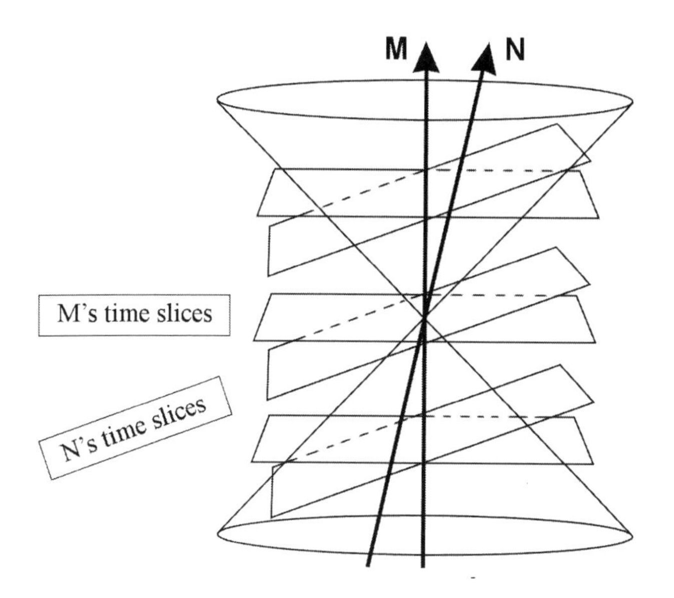

*Figure 3.4: Slices of simultaneity according to observers M and N, whose world lines are **M** and **N** respectively.*

In Figure 3.4, M's world line **M** is vertically upward. M's time slices are horizontal sheets.

N's world line **N** is tilted to the right as shown. N's time slices are tilted upwards in the direction of his travel with respect to M. (But, in the direction into and out of the page, these sheets are horizontal.) These respective tilts – of N's world line and time slices – are equal is size (this is seen more easily in Figure 3.3).

Suppose N increases their velocity relative to M. What changes would this bring about in the diagram? The world line **N** would tilt over more, and N's time slices would tip up by the same amount.

A particle travelling very close to the speed of light, as measured in our reference frame, must have both their world line and their time slices extremely distorted, being tilted extremely close to the light cone. But remember, we are free to redraw the diagram in the rest frame of this particle. In this new frame, the particle's world line is vertical, and its time slices horizontal. In this new situation, our world line and time slices are highly tilted, but in the opposite direction.

The Twin's Paradox

This is a scenario in which twin Ann makes a rocket trip to a star, and immediately returns home. Her spaceship travels with speed v, as measured by twin Bob who stays home. Because of time dilation, Ann will have aged less than Bob when they meet up again in the debriefing room. If v is close to c, the discrepancy can be extreme: Bob might be an old man, while Ann has hardly aged at all. This situation, whilst bizarre, is logically consistent, and is the way in which the world actually works according to special relativity.

From Ann's viewpoint, Bob is travelling with speed v relative to her at all times. The alleged paradox arises if we **wrongly assume** that *only the relative velocities of Ann and Bob are important*. With this assumption, by the argument just given, Bob must have aged much less than Ann: Bob is still young at the end of the trip, whereas Ann is an elderly woman.

This is a real contradiction. It cannot be true that each twin is many years older than the other when they are chatting in the debriefing room. The paradox is resolved when we realise that the situation is not symmetric: homebody Bob remains in the same inertial reference frame throughout; whereas Ann must change her inertial frame whenever she fires her rocket; most importantly to turn it around upon reaching the star. The italicised statement in the previous paragraph is false.

This is all that needs to be said, but I wish to give a detailed example in which the numbers have been chosen for easy calculation. The situation is summarised in Figure 3.5.

In this Figure, V is the event that the rocket is launched,

and Z is the event that it lands back on Earth. S is the event when the rocket turns round at the star. Event X is the point on Bob's world line which, in his reference frame, is simultaneous with S: According to Bob, event X happens just as Ann reaches the star.

Suppose the star is 10 lightyears away, as represented by the unmarked line XS.

Suppose Ann travels to the star at speed $v = 0.6c$. For simplicity we will assume that the rocket is extremely powerful and reaches this velocity in a negligible time. This latter assumption, while unrealistic, does not affect the validity of the Twin's Paradox argument. At this speed the Lorentz factor $\gamma = 5/4$, as you may check.

Bob's timeline: By Bob's reckoning: The time Ann takes to reach the star is equal to its distance, divided by the velocity of the spaceship. So, the flight time $f = 10 / 0.6 = 16\frac{2}{3}$ years according to Bob. This is represented by the line VX in the Figure. The round-trip time is twice this: $2f = 33\frac{1}{3}$ years, represented by the length of the vertical line VZ.

Ann's trip: As measured by Bob, Ann's clock's run slower by a factor γ. Therefore, according to Ann in the spaceship reference frame, her flight time f' to the star is given by $f' = f / \gamma = 16\frac{2}{3} \times (4/5) = 13\frac{1}{3}$ years. Her round-trip travel time is $2f' = 26\frac{2}{3}$ years.

(We can derive f' in an alternative way, working entirely in Ann's reference frame. So far as she is concerned, her clocks are running perfectly normally. But, as soon as her voyage is under way, she notices that star S is now closer by a factor γ, because of Lorentz contraction.)

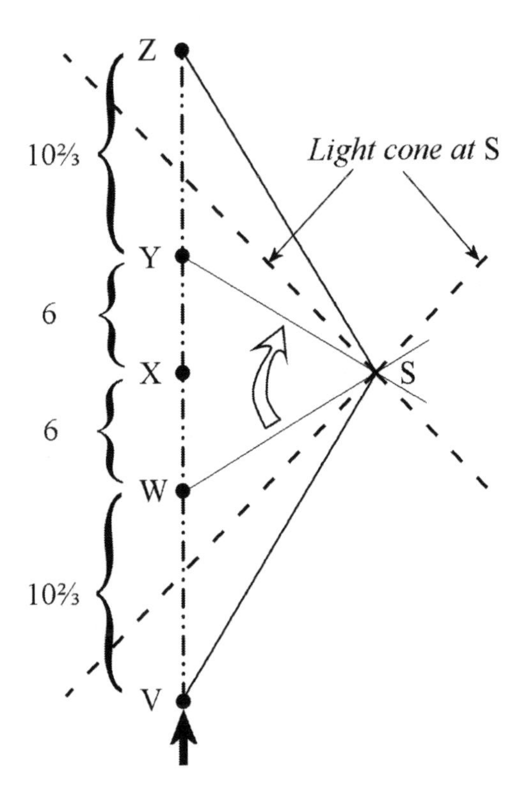

Figure 3.5: The Twin's Paradox. Bob stays at home: his world line is VWXYZ. His twin Ann travels to star S, which is 10 lightyears away, at velocity $v = 0.6c$. She returns home at the same speed: her world line is VSZ.

According to special relativity, when the twins come together in the debriefing room, Ann is now younger than Bob by 2f – 2f′ = 6⅔ years – a substantial amount!

Explanation: On the outward journey, Ann similarly deduces that Bob's clocks are running slower by a factor γ. As Ann nears the star, her clock reads f′, so at that moment Bob's clock must, according to her, read $f' / \gamma = 13\frac{1}{3} \times (4/5) = 10\frac{2}{3}$ years. This means that line WS in Figure 3.5 is a slice of simultaneity in Ann's reference frame on the approach to S.

After Ann has completed firing her rocket at S, her speed is a constant $v = 0.6c$, but now directed back towards home. It can be shown by a similar argument that event Y on Bob's world line is such that line YS in the Figure is a slice of simultaneity in Ann's new frame for the return trip.

What happens in the interval WY, according to Bob and Ann? Bob's observation must be that 12 years have passed uneventfully. According to Ann, however, she has been accelerating while her spaceship is firing for a short while, say just for an hour. During this interval she is forced back in her seat, and her reference frame is continuously changing: starting from the outbound frame, and finishing at the inbound one. In particular, her slices of simultaneity have rotated, from WS to YS, as indicated by the large arrow in the Figure.

In this brief period of acceleration, Ann's clock has changed by 1 hour, whereas she deduces that Bob's clock has advanced by 12 years, as shown by line WY. Ann has measured a moving clock that is ticking with incredible rapidity. This does not contradict the time dilation rule of special relativity because – while the rocket is firing – Ann does not remain within any single inertial reference frame.

Solving problems

Many problems in relativity can be solved in a similar manner:

First, always remember to attribute every measurement to a particular inertial frame. For example, which of W and X is simultaneous with event S? As it stands, this is not a meaningful question. W and S are simultaneous in the frame of Ann's outward journey. X and S are simultaneous in Bob's frame.

Second, given information about happenings in frame R, use the Lorentz transformation to find out what is going on in frame R′.

General relativity

During 1907-15, Einstein developed his theory of *general relativity*, which extended special relativity, and was his explanation of matter and gravitation in terms of the curvature of both space and time. In mathematical terms, *curvature* means that the formula for the spacetime interval I(A, B) is more complicated than the one given previously for flat Minkowski spacetime. One consequence is that light cones – defined by I(A, B) = 0 – can no longer be lined up perfectly parallel to one another, as in Figure 3.2; instead they must be bent over in a way specified by the new formula.

Large masses, such as the Sun, cause a significant curvature in spacetime. Comparatively light objects, such as planets, travel along the straightest possible paths, called *geodesics*, in spacetime.

A comprehensive university text introduces general relativity with "The parable of the apple." After attending their first lecture, a student sees an ant crawling over the surface of an apple, and makes the following comparisons: The stalk is analogous to the massive Sun. The curvature of the apple's surface near the stalk is like the curvature of spacetime caused by the Sun. An ant, turning neither left nor right, and following the straightest possible path over the surface of the apple (a geodesic), will be deflected in such a way that the student sees that its path necessarily curves towards the stalk. The student sums up:

> *Space acts on matter, telling it how to move. In turn, matter reacts back on space, telling it how to curve.*
> (Original italics, Misner, Thorne & Wheeler, 1973, pp. 1-5)

Sufficiently small patches of spacetime look flat – like Minkowski space. The approximation gets better as the patch gets smaller. This is akin to examining a few square millimetres of the apple's surface with a magnifying glass and seeing that this is approximately flat. Another example is that a flat map of the UK can represent this country with little distortion, because it is a small region on the surface of the spherical Earth. This is in contrast with Africa which, because it covers a large portion of the Earth's surface, cannot be depicted on a single map without significant distortion.

As with any parable, there are some disanalogies. Two of the most important are: (1) In the parable, time is treated separately, whereas in general relativity time is fully integrated into a single entity, spacetime; (2) In the parable, the masses (the ant and the stalk) exist in addition to the apple's surface. But, in general relativity, matter is identical to a particular type of spacetime curvature – it is not anything extra. The world line of the Earth is more akin to a timelike crinkle in spacetime, rather than to an entity 'moving over the surface' of spacetime.

Principle of equivalence

A major concept that enabled Einstein to develop his general theory was his *principle of equivalence*: It is impossible to perform any experiment within a sealed room that will enable you to detect whether it is static within a gravitational field or whether it is in empty space being accelerated upwards by an equivalent force. For example, it is impossible to tell if you are standing in a static lift, or if you are standing within a rocket in outer space being accelerated at 1g (= 9.8 metres per sec^2). Similarly, if you suddenly begin to float around the cabin, it is impossible to tell if the cable supporting the lift has snapped, or if the rocket's engine has been switched off. Astronauts float freely in their spacecraft while it is in orbit about the Earth. The spacecraft is analogous to the freely falling lift, and the astronauts are like the lift's passengers.

Experimental verification

General relativity laws achieved substantial experimental

verification. On its first publication, Einstein was able to account for a small anomaly in the orbit of Mercury. For centuries this had remained inexplicable in terms of Newton's gravitational theory.

General relativity also predicted that light rays would be bent slightly as they passed through the gravitational field of the Sun (because light cones were very slightly tipped towards the Sun). Einstein predicted the size of this effect (giving a figure twice that of Newton's theory). Because of World War I, an experimental test was delayed but, in 1919, Arthur Eddington led an expedition to measure star positions, apparently displaced by the gravitational field of the Sun. The measurements had to be made during a total eclipse, because at all other times light from the Sun obscures light from the stars. This test sealed Einstein's fame, both for scientists, and in the public eye.

Other confirmations came much later. Einstein used the principle of equivalence to predict that clocks in a strong gravitational field would run slower than those in a weaker field (1920, ch. XXIII). On the surface of the Earth, for example, clocks run more slowly than identical clocks carried on orbiting satellites. In this example, the effect is only of the order of microseconds per day but – in confirmation of Einstein – it needs to be taken into account if GPS systems are to work correctly.

Einstein predicted the existence of *black holes* – regions of spacetime in which gravitational effects were so strong that not even light can escape. Anything falling into a black hole would be ripped apart by extreme tidal forces before hitting the singularity at which the laws of physics break down. Einstein himself believed that black holes were merely theoretical – they were just mathematical artefacts that could not exist in nature. Later work has shown that stars bigger than about three solar masses (these are commonplace) must collapse into black holes when their nuclear fuel is spent. There is also excellent evidence for a supermassive black hole (of about 4 million solar masses) at the centre of our own Milky Way galaxy.

One prediction of Einstein was that, when massive objects orbit one another, *gravitational waves* – minute ripples in spacetime – would be emitted. These were not observed directly until 2015, when two black holes orbiting each other were witnessed to have merged into one. The instruments that achieved this feat have to be incredibly sensitive – measuring motions 10,000 times smaller than an atomic nucleus (see https://www.ligo.caltech.edu/).

Cosmological models

General relativity was the beginning of *cosmology* – the study of the universe as a whole – as a true science. Of necessity, cosmological models are highly simplified. Typically, they assume a **uniform** distribution of matter throughout space, having the same density as the **average** density of the actual matter in the universe. But, as you can confirm by looking at the night sky, the actual distribution of matter is decidedly lumpy.

Most cosmological models are highly symmetric. Because of this there is a preferred spatial reference frame that is static with regard to average distribution of matter. There is also a preferred global concept of cosmic time. This is defined by the passage of time along the world lines of each piece of matter (say each speck of dust).

Einstein's equations have a certain amount of flexibility: there is a *cosmological constant* Λ that can be chosen arbitrarily, yet still giving valid, alternative solutions to his equations. This Λ was chosen by Einstein so as to give a model universe that was static. Prior to his time, throughout human history, everyone believed that our universe was static. In 1929 Edwin Hubble published results that suggested galaxies were receding from us, with a velocity proportional to their distance from us. Of course, it is implausible that we are at a special central position in the cosmos. By setting $\Lambda = 0$, Einstein was able to create models that were consistent with Hubble's evidence, but in which the universe is expanding everywhere, from an initial, extremely compressed state, which occurred about (to give the current figure) 13.7 billion years ago. The initial paroxysm later came to be called the "Big Bang". Einstein deemed his non-zero Λ to be, "The biggest blunder of my life." Despite this, the cosmological constant occasionally reappears in cosmological models (although not of course in order to revert to a static universe).

Spacelike slices (i.e., 3-D slices at any given cosmic time) within Einstein's models have positive ('sphere like'), zero ('flat'), or negative ('saddle like') curvature. Universes with positive curvature eventually collapse under gravitation to a point: a final 'Big Crunch' that mirrors the initial Big Bang. Those with flat spacelike slices will avoid collapse – but only just. Universes with saddle-like slices expand more vigorously. The actual curvature of spacelike slices can be measured from astronomical observations.

Determinism

Special theory is deterministic; and the causal structure of general relativity is known to be similar. In general relativity, determinism has not yet been **proven** to be valid without exception, though this is believed to be the case. There might be (albeit implausible) exceptions in extreme circumstances, such as near black holes; Penrose discusses this (1989, pp. 273-8).

Communalities

The term *classical* is fairly recent, and was coined in order to contrast the theories above with the new quantum theories, to be discussed in the following chapters. Here are some features which classical theories share in common:

- *Realism*

The universe exists and goes about its customary business largely independently of us. In particular, it existed long before the coming into being of any life on Earth, and it will continue to exist after the extinction of humanity, and indeed after the extinction of life everywhere.

- *Measurement*

When measuring any physical property (say the flow of a stream), the measuring process inevitably causes some disturbance to the object being measured. But this disturbance can be made as small as we wish by making the interaction smaller. We could for instance measure the flow using a miniscule waterwheel. In classical theories we can, in principle, make measurements to any desired degree of accuracy. For the same reason the property being measured can safely be assumed to exist objectively.

- *Universal laws*

The dynamical development of the cosmos is governed by universal mathematical laws, which means that they apply at all places and at all times – without exception. These laws are objective facts of our universe, which is why we talk of *discovering* (rather than 'inventing') them.

Of course, scientific theories are human creations. For a period after their first publication, they may be taken as truth, but it always remains possible that they will subsequently be found to be incorrect. Newton's theory in particular is historically notable for being taken to be the absolute, ultimate truth for many centuries and for its eventual falsification. Nonetheless, throughout the classical period it was firmly held that our universe is governed by objectively true laws, of which we can only have partial and provisional knowledge.

- *Determinism*

Classical theories are deterministic. In principle, the totality of facts about the precise state of the universe at any given time would allow all future states to be predicted exactly.

As discussed earlier, classical thermodynamics is a statistical theory, but in it the uncertainty arises solely as a result of our lack of complete knowledge. If we were somehow magically given this knowledge, it would become apparent that underpinning this theory is deterministic Newtonian mechanics.

- *Locality*

Locality was such a natural assumption during the era of classical science (which lasted from the sixteenth to the beginning of the twentieth century) that it was never formally stated. If we wish to express it now, it would be something like **L**: "Goings-on in one small region (call it A) of the universe cannot instantly or significantly **change** what is going on in another distant small region (B)."

Our initial intuition about condition **L** is that, if it is false, it would render science impossible. Imagine a scientist setting up an experiment in region B. Goings-on in distant region A could alter the results of this experiment. Moreover, there would be an infinite number of different faraway regions, each creating their own disturbance. As we shall see in chapter 5, quantum mechanics **is** non-local, but this non-locality is so subtle that it does not reduce science to chaos – scientists can still make predictions of exquisite accuracy.

According to Newton's theory, waving my fist instantly causes the planet Mars to jiggle in its orbit. This jiggling is utterly insignificant – it is far too small to alter any experiment taking place on Mars. Newton's theory therefore satisfies **L**.

Einstein's theories are strictly local (provided one takes

the theories as they were originally presented, and does not attempt to add extraneous quantum ideas). Both special and general relativity satisfy **L**, even if we remove the clause "or significantly" – in other words, for these theories the amount of instantaneous, non-local change is exactly zero. We can also simplify the expression for **L** by stating, "If small regions A and B are spacelike separated..."

- *Familiarity*

In classical theories all entities in the universe are, in general terms, considered to be analogous to those entities we experience, and are thus familiar with, in everyday life. For example, a planet, a billiard ball, a peppercorn, a pollen grain, and an atom can all be regarded as spheres of differing sizes and masses, and having a definite location at each instant of time. All are analogous to their most commonplace representative (the ball). Differences among them can be explained in terms of their differing properties in a clearly understandable way. For example, pollen grains are so small and light that they float, and are buffeted around by air molecules.

- *The task of science*

The task of classical science is to understand our universe, and to discover the laws governing its existence.

Suppose we have a recipe book that tells scientists what to do in order to set up apparatus and carry out an experiment – including what measurements to make and record. Suppose further that the recipe instructs scientists what mathematical procedures to follow in order to predict these experimental results with superb accuracy. By classical lights, such a recipe book would not amount to a scientific theory because – of itself – it gives no understanding of our cosmos.

Chapter 4

Quantum mechanics

What is quantum mechanics?

Here is a brief sketch of some of the theory's main features:

- *Quantum mechanics is a theory that deals with entities that are beyond our direct experience*

Such entities include atoms, electrons, and so on (rather than perceptible things like clouds, items of apparatus, and birds).

- *Quantum mechanics obtains knowledge about these unexperienced entities by their effects on what is accessible to human perception*

To give some examples of such effects: the audible click of a Geiger counter; a visible mark on a photographic plate; and data from the Large Hadron Collider, whether seen on a computer screen or in a printout. The scope of quantum theory is vast: it explains the processes involved in the interiors of stars; the stability of atoms over huge expanses of time; computer-related technologies; and much else.

- *In terms of the* ***accuracy of its predictions****, quantum mechanics is by far the most successful theory of all time*

It predicts the results of some experiments with an accuracy of one part in a thousand trillion (10^{15}). For example, Parthey, Matveev, and 15 others measured the frequency of light in a particular line in the spectrum of a hydrogen atom to be 2,466,061,413,187,035 Hz, with an accuracy of ±10 Hz (2011).

- *Quantum mechanics is beyond weird*

Weirdness alone is not enough to call a physical theory into question – we are living in a strange universe. But with quantum mechanics the weirdness is so extreme that its coherence and meaning are disputed. This may be contrasted with general relativity, whose effects, although bizarre, and unfamiliar, are readily seen to be logically self-consistent to those familiar with the mathematics. Moreover, relativity gives an unproblematic picture of what the world is like, and of the goings-on within it. In contrast, proponents of the dominant school of thought in quantum mechanics hold that the defining goal of physicists – the endeavour to discover what the stuff of the world is actually doing – should be abandoned. For many decades they actively denigrated fellow scientists who continued in this task; this will be the topic of the next chapter.

- *In quantum mechanics, measurement is taken to be basic, and hence unexplained*

This is an extremely problematic feature of the theory, which makes it hard to find credible. In one sense, it is perfectly clear to any experimenter when a measurement happens – a Geiger counter clicks, an instrument pointer moves to the right, and so on – but, for a viable physical theory, one should have the ambition to give an account of measurement **in physical terms**; in other words, to give an account of apparatus, and indeed scientists, in terms of their constituent atoms (and similar micro-entities). Quantum mechanics does not provide these.

How to begin?

There are formidable problems in introducing quantum mechanics. One might take a historical approach, but the theory came into being in an extremely messy, confusing and piecemeal way over several decades. This history is far from finished and the battle over ideas is not yet won.

Another possibility is to take each single experimental situation in turn, and show how the machinery of quantum mechanics predicts the correct results. The difficulty with this approach is that the machinery appears to be hugely excessive to explain any single experiment: The complicated mathematics is necessary in order to give a comprehensive account of every quantum experiment that has been performed in the past century.

Tim Maudlin's procedure

Tim Maudlin has developed a powerful method for minimising confusion, which I will follow in this book. In *Philosophy of Physics: Quantum Theory* (2019), he first discusses the physical facts that the theory is attempting to explain. He does this by describing a representative selection of experiments and the results one obtains. As far as possible this discussion avoids any attempt at theorising. (In describing experiments, one is bound to bring in certain physical concepts, such as 'electron'; but the intention is to avoid any commitment as to what sort of thing an electron might be.) All that is of interest at this stage is in what the world actually does.

Next, he introduces what he calls "the quantum recipe", a set of mathematical rules and procedures that, as he demonstrates, will correctly predict the results of all of the experiments that he has described. He emphasises that quantum mechanics **does not amount to a physical theory** because it says nothing about what is going on in the system under consideration – it is merely a recipe book for correctly predicting the outcomes of experiments. This is an important insight, and explains why bare quantum mechanics, unlike other physical theories, uniquely requires an interpretation. The purpose of an *interpretation* is to add whatsoever extra is necessary to bare quantum mechanics, in order to convert it into a proper *physical theory*: an account of what exists in the world, and what is happening to it.

The final step is to discuss a representative sample of interpretations. (Maudlin does not cover the "Copenhagen interpretation", except to criticise it in passing, because it amounts to accepting bare quantum mechanics (i.e., recipes). Its advocates stoutly reject any attempt at interpretation.)

This threefold separation – **(a)** Collect facts about the world; **(b)** Use quantum mechanics to correctly predict these facts; **(c)** Attempt to understand the world by providing an interpretation – is superbly useful in clarifying the issues to be resolved. With hindsight these can readily be discussed independently and in the correct logical order. (Historically this was far from being the case.)

Once pointed out, it is clear that quantum mechanics without an interpretation is a methodology for correctly predicting experimental results, and does not of itself amount to a scientific theory. Moreover, this is why it requires an interpretation; and why no other theories do – two facts that are puzzling when one first encounters quantum theory.

Representative experiments and their results

Maudlin picks out representative experiments and their results. They are all well-known and are standardly described in many textbooks.

Young's Slits experiments

Figure 4.1 shows schematically the setup of a very famous group of experiments. To the left is a *source* of particles of known momentum. In the centre is a *barrier* with two narrow slits, very close together. On the right is a *target* to detect the particles.

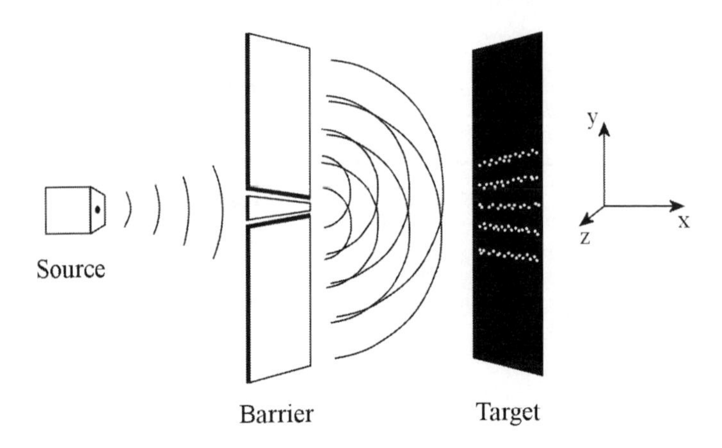

Figure 4.1: The Young's Slits experiment. Also shown (displaced for clarity) is a coordinate system (which really has its origin between the slits).

When only one slit is open, the particle beam travels in a more-or-less straight line from the source, through the slit to the target. Figure 4.2(a) shows the pattern of intensity of the beam across the target. This pattern shows that the beam has spread out after passing through the slit. (The spreading is greater when the slit is very narrow: For a wide slit, the beam does not spread significantly.) Similarly, if the other slit had been open then the same pattern of intensity would be seen; but slightly displaced, so as to be directly opposite the now-open slit, as shown in the other curve in the Figure.

When both slits are open, then the pattern of intensity at the target has the form of a series of bands, called *interference fringes*, as shown in Figure 4.2(b). (These bands are also sketched in Figure 4.1.)

How are we to interpret these experiments? When Thomas Young first performed these experiments – with light – in 1801, they were taken to be evidence that light has the character of a wave. Here is an analogue model: In Figure 4.1 we can imagine looking vertically down on a tray of water. Instead of source, we have a block moving up and down creating waves, whose crests are sketched in the Figure. Waves pass through the slits in the barrier, and begin diverging in semicircles from each slit.

In the right side of the tray there is a complex pattern of ripples. Where the crests of waves coming from both slits coincide, the water is highest. Where the troughs of the waves coming from both slits coincide, the water is at its lowest. Where the crest of the wave coming from one slit coincides with the trough of the wave coming from the other, then the water will be at its natural, resting level. The result is that the pattern of the intensity of these water waves at the target is very similar to the interference fringes of Figure 4.2(b).

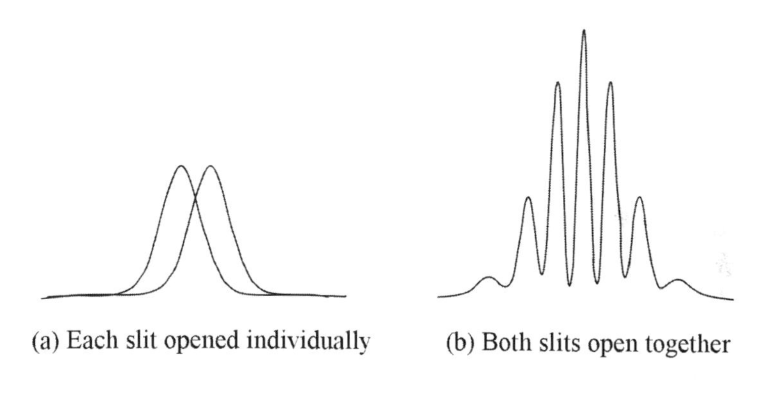

*Figure 4.2: Young's Slits – the pattern of light intensity across the target when **(a)** each slit is opened with the other remaining closed; **(b)** both slits are open together.*

Another, more general, argument from this era in favour of the wave theory of light is that **less light** arrives at certain places on the target when both slits are open, as compared to when one slit is open. This strongly suggests that something non-localised is coming through both slits.

In the twentieth century it was possible to refine the experiment sufficiently to be able to detect individual scintillations arriving at the target; this is strong evidence that light has particle-like characteristics, and such particles of light are called *photons*. Moreover, because the speed of light is known, it is possible to reduce the intensity of the light to such an extent that there is only one photon within the apparatus at any given time. This would be the case, for instance, if the average rate of scintillations that occur is only a few per second. Despite this situation, the same pattern of interference fringes builds up.

A variant of the Young's Slits experiment is to add some appropriate apparatus in order to observe which slit each photon passes through. If this is done, then, no matter what apparatus is used, the interference fringes are lost. The intensity pattern at the target changes to the sum of the intensity patterns for each individual slit when opened separately; i.e., to the sum of the two curves depicted in Figure 4.2(a). This latter intensity pattern is a broad curve having a single smooth peak.

Light thus has seemingly contradictory wave-like and particle-like characteristics. This is merely a statement of the problem of the nature of light – it is not intended to amount to a solution. We loosely refer to light – or any other entity – as a 'particle' when it displays particle-like characteristics (notably, that of being highly localised), and as a 'wave' when it exhibits wave-like characteristics (notably, that of being spread out). Such entities form a 'beam' when they follow a particular route through the experimental setup. I will continue to use such terms freely, but it must be remembered that they are not to be taken literally, and can be extremely misleading.

The results of the Young's Slits experiments are extremely peculiar, but perhaps photons are stranger than we first imagined. Even weirder is that essentially the same experiments have been performed, with identical results, using large molecules. The most well-known experiment was performed with Buckminsterfullerene (a molecule consisting of 60 carbon atoms arranged in a sphere – rather like a football); but it has since been performed with even larger molecules containing over 800 atoms. This is utterly astonishing – especially to those of us whose strongly ingrained physical concept, as to what a 'molecule' is really like, is similar to that of a stick-and-ball model. (We have just used such a model to describe the structure of Buckminsterfullerene – it would be incredibly difficult to form any concept of the molecule without it.)

A point to be noted about this experiment – but it is one that invariably applies to all experiments in quantum mechanics – is that the particle-like character of entities is more-or-less directly observable in terms of position measurements, such as the scintillations here. In contrast, the wave-like character of these entities is suggested at one remove – only being inferred, after much mathematical theorising, on the basis of a multitude of position measurements. Unlike water waves, in quantum mechanics, wave-like behaviour is never seen directly. The wave crests sketched in Figure 4.1, for example, are inferred theoretically from the scintillations observed on the Target. (See Maudlin, 2019, pp. 8-17.)

Basic spin experiments

Figure 4.3 is a simplified diagram showing a Stern-Gerlach apparatus (named after its inventors). It measures something that physicists call *spin*, though this has very little to do with spin as we are familiar with it in our everyday lives.

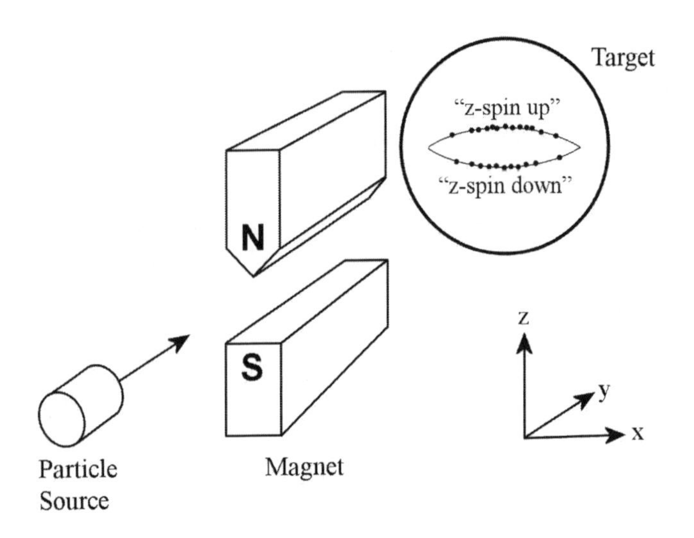

Figure 4.3: The Stern-Gerlach apparatus

The Stern-Gerlach apparatus (S-G for short) consists of a specially shaped magnet with the poles very close together. Only the poles are depicted: the remainder of the magnet is not shown. There is a magnetic field between the poles of the magnet, and, because the north pole has a pointed shape, the magnetic field is stronger near it. Arriving from one end is a source of particles, which travel through the gap between the poles of the magnet and ultimately land on a target – for example a photographic plate. The place the particle lands can be regarded as providing the measurement of spin.

Throughout this book, whenever we discuss spin, it will always be in reference to so-called *spin-½* ('spin half') particles, which include protons, neutrons, and electrons. With such particles, the output is as given in the diagram: half the particles are deflected 'up' towards the north pole of the magnet; and half are deflected 'down' towards the south pole of the magnet. (By convention, 'up' is **always** towards the pointed north pole.)

You will notice that the particles that did not follow the

exact mid-line (given by x = 0) through the apparatus were less deflected up or down; and those far away from the mid-line were not deflected at all. The classical explanation would be that particles going into the apparatus askew spent less time in the non-uniform magnetic field.

This experimentally observed pattern of deflections on the target – empty in the middle – is unexpected: What classical electromagnetic theory predicts is that this canoe-shaped outline should also be filled-in with scintillations.

We can also set up a coordinate system in the laboratory, in order to discuss the layout of the apparatus. As shown in the Figure, the beam of particles is travelling in the y direction, and the Stern-Gerlach apparatus is oriented in the z direction. In this situation, we say that the S-G apparatus is oriented so as to measure 'z-spin', and the two possible outcomes are 'z-spin up' and 'z-spin down'. It is easy to see that, by rotating the S-G apparatus 90° clockwise, we could instead measure the spin of each particle in the x direction.

Just for the record, here are a few additional experimental facts about spin: Spin always occurs in half-integer values, that is to say, 0, ½, 1, 1½, and so on. The amount of spin possessed by any particle of a given type is always fixed in size. For instance, all electrons have spin-½ – it is impossible for an electron to 'spin twice as fast', whatever that might mean.

Now let us describe some experiments. Replace the target with a barrier having a slot so that only z-spin up particles can go through. This is called *'preparing a beam of z-spin up particles'*. Measure these prepared particles with a second S-G apparatus oriented in the same direction. We find that all of these particles are still z-spin up. (Likewise, z-spin down particles are always z-spin down when measured for a second time.)

Now, instead of taking our prepared z-spin up particles and measuring them in the z direction, let us measure them in the x direction instead. We find that half are x-spin up, and half are x-spin down. So, knowing the z-spin tells us nothing about the x-spin.

To sum up so far: Prepare a beam of z-spin up particles; if we measure in the z direction again, then we have perfect foreknowledge of the second measurement; if we instead measure in the direction x perpendicular to z then we have zero foreknowledge. This suggests that we try the experiments of rotating the S-G apparatus through an angle θ (in the x-z plane), and find out the probability of getting *agreement*, defined as both pieces of apparatus reading 'up'. We can repeat this experiment for any value of θ that we wish. When these experiments are performed, it is found that the probability of agreement is $\cos^2(\theta/2)$.

(Here is a quick check that this formula – which is derived from experimental results – is consistent with our two examples: When $\theta = 0$, then clearly $\cos^2(0) = 1$ as required. When $\theta = 90^\circ$, $\cos(\theta/2) = 1/\sqrt{2}$, so $\cos^2(\theta/2) = ½$ as required.)

(Another aside, relevant only to those who read widely: Despite their seeming differences, $\cos^2(\theta/2)$ is mathematically identical to $(1 + \cos(\theta))/2$, and some authors prefer to use the latter expression.)

Lastly, suppose we prepare a beam of z-spin up particles. We next measure their spin in the x direction, and finally measure their spin in the z direction again. Do we find them all still to be z-spin up? No! Only half of these particles are now z-spin up. It seems – at least at first sight – that in measuring spin in the x direction, all previous information about spin in the z direction has been lost. (Contrast this with measuring z-spin twice in succession.) (See Maudlin, 2019, pp. 17-22.)

Interferometer experiments

Figure 4.4 shows a Mach-Zehnder interferometer, named after Ludwig Mach and Ludwig Zehnder, who invented it in 1891.

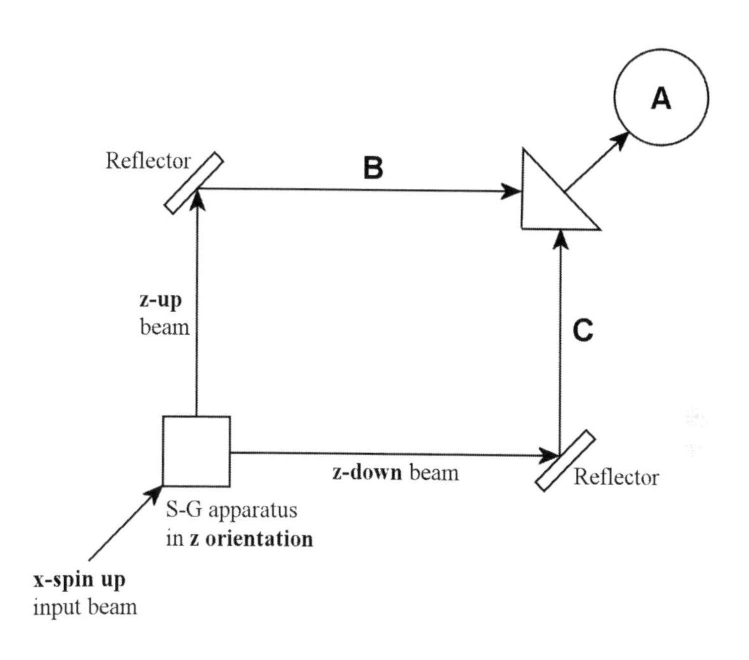

Figure 4.4: The interferometer

The general principle is to have a suitably prepared input beam enter the interferometer from a source; the beam is split into two physically distinguishable parts, which take widely separated routes through the interferometer. These beams are then recombined into one, and this is measured by some further apparatus at point **A**.

The particular setup considered by Maudlin is shown in the Figure. Electrons, which are spin-½ particles, are prepared in an x-spin up state, and these enter the interferometer at the lower left. Here, a z-oriented S-G apparatus divides the beam into two physically distinct halves: one beam is z-spin up, and the other beam is z-spin down. At this lower left location, additional equipment (not shown) then directs these output beams along paths that are at right angles to one another.

The z-spin down beam takes the lower route through the

interferometer, and the z-spin up beam takes the upper route. These beams are first reflected, and are then reunited by the device, symbolised by a triangle, at the top-right of the Figure. This single beam then exits the interferometer.

The first experiment with this interferometer is to measure the x-spin of the recombined beam at location **A** by placing an x-oriented S-G apparatus there. When this is done, all the particles are found to be x-spin up.

This experimental result should astonish you. Recall the final experiment in the ***Basic spin experiments*** subsection: when we prepare an x-spin up beam, and then measure its z-spin, the two output beams, z-spin up and z-spin down, **each considered in isolation**, appear to contain no information about x-spin. This may be confirmed by placing x-oriented S-G apparatuses at locations **B** and **C**. At each of these locations, half of the particles will be found to be x-spin up, and half x-spin down. So, on the evidence of the two beams considered individually, all of the information about x-spin **appears** to have been destroyed. Somehow, the two widely separated sub-beams between them hold the information about x-spin, even though neither portion, considered in isolation, appeared to do so.

Likewise, if a beam of x-spin down is input into this device, the output beam will be entirely x-spin down. Moreover, as with the Young's Slits experiment, if the intensity of the beam is turned down sufficiently to ensure that only one electron is within the device at a time, the experimental outcomes remain unaltered.

The second experiment that can be performed with the interferometer is very subtle. There is a device, called a "do nothing box", which in most situations has no effect whatsoever. For example, all of the experiments in the previous ***Basic spin experiments*** subsection would give exactly the same results – whether or not this device was present. Maudlin (2019, pp. 24-25) gives slightly more detail. For present purposes it is sufficient to note that the device contains a specific magnetic field. The special situations in which the device **does** have physical consequences are those in which it is inserted into one portion of a beam that has been divided, and in which all portions of this beam are subsequently reunited.

We repeat the first experiment, but on this occasion inserting the "do nothing box" into the interferometer at **B**. Now when we input x-spin up particles into the interferometer, they exit with the opposite spin, namely x-spin down. Similarly, inputting x-spin down particles results in x-spin up particles on exit. The same results would have occurred if we had placed the "do nothing box" at location **C** instead. (See Maudlin, 2019, pp. 22-25.)

The quantum recipe: part I, maths

We now wish to introduce the rules and procedures of quantum mechanics that enable correct predictions to be made. This involves quite extensive mathematical knowledge. It is impossible within a reasonable space to explain everything, and I will not attempt to do so. Recall that such all-encompassing machinery is required because we want to provide a **single procedure** that will produce correct predictions in **all** relevant experimental situations, including the representative examples above.

I'm sure you are familiar with ***real numbers***, such as 1, -2.75, and π; and the operations that can be performed on them: +, -, ×, ÷, √, etc. The multiplication sign is usually omitted, so $2\times\pi$ is written 2π. The real numbers have a natural ordering ('<') because they can be depicted along a line.

Important to the theory are ***complex numbers***. These follow all the usual rules of arithmetic, except we have a new number, i, with the property that $i^2 = -1$. All complex numbers, c, can be written in the form $c = a + ib$, where both a and b are real numbers. Moreover, c can be represented on a plane as a point with coordinates (a, b).

Because complex numbers live on a plane, they do not have

any natural ordering. For the same reason they can also be expressed in polar coordinates: c = (r, θ), where r is the *amplitude* (distance from the origin), and θ is a *phase* (angle measured around from the horizontal axis). Both r and θ are real numbers. These polar coordinates are usually written $c = re^{i\theta}$ (where *e* is the mathematical constant 2.718...). This can also be written $c = r\cos(\theta) + i\, r\sin(\theta)$.

If c is the complex number a + *i*b, then it is related to another, denoted c*, obtained by reversing the sign of b. This c* = a – *i*b is called the *complex conjugate* of c. Multiplying a complex number by its complex conjugate gives the square of its amplitude:

$$cc^* = (a + ib)(a - ib) = a^2 - iab + iab - i^2b^2 = a^2 + b^2 = r^2$$

The final step follows by Pythagoras' Theorem.

Next consider ***real vector spaces***. Let us take three dimensions to be specific. Non-zero vectors can be represented by arrows beginning at the origin (0, 0, 0) and ending with its tip at some other point, say with coordinates (1, -2, 4.6). We may identify this vector, **v** say, with the coordinates of its tip, so we can write **v** = (1, -2, 4.6).

Let **w** = (2, 6, -5) be another 3-D vector. We can add **v** to **w** by adding their coordinates:

$$\mathbf{v} + \mathbf{w} = (1 + 2, -2 + 6, 4.6 + -5) = (3, 4, -0.4)$$

To visualise this, we can imagine **v** and **w** as two adjacent sides of a parallelogram, and complete the figure by drawing the other two sides. Then **v** + **w** is the particular diagonal of the parallelogram that begins at the origin.

In a real vector space, the real numbers are called scalars. To multiply **v** by a scalar (say 2) you simply multiply each of its coordinates by 2:

$2\mathbf{v} = 2(1, -2, 4.6) = (2, -4, 9.2)$

Geometrically, arrow $2\mathbf{v}$ is twice the length of $\mathbf{v}$ (which is probably where the term 'scalar' came from). Multiplying $\mathbf{v}$ by the scalar -1 reverses the signs of all $\mathbf{v}$'s components:

$-1\mathbf{v} = -\mathbf{v} = (-1, 2, -4.6)$

Vector addition and multiplication by scalars are pictured in Figure 4.5.

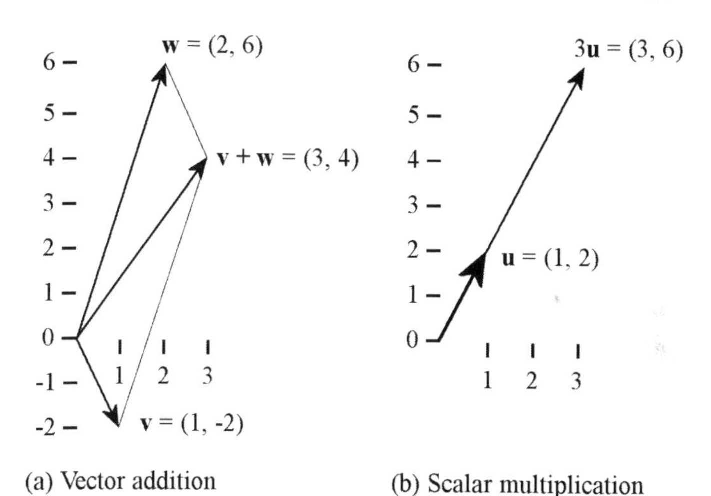

Figure 4.5: Vectors: ***(a)*** *vector addition can be done using the parallelogram rule, or by adding corresponding components;* ***(b)*** *multiplying by a scalar (e.g. 3.0) can be done by stretching the vector by the scalar, or by multiplying each component by the scalar.*

There is another useful tool – which is not part of the definition of a vector space – that helps us to find the lengths of and angles between vectors. It is called the *'dot product'* or *'inner product'* of two vectors. It is a scalar, calculated by multiplying the corresponding coordinates of a pair of vectors together and then adding. Take **v** and **w** as before:

$$\mathbf{v} \cdot \mathbf{w} = (1, -2, 4.6) \cdot (2, 6, -5)$$

$$= 1\times2 + (-2)\times6 + 4.6\times(-5)$$

$$= 2 - 12 - 23 = -33$$

Define the *norm* ('length') of a vector **v** to be the positive number:

$$\|\mathbf{v}\| = \sqrt{(\mathbf{v} \cdot \mathbf{v})}$$

Because $\mathbf{v} \cdot \mathbf{v} = (1, -2, 4.6) \cdot (1, -2, 4.6) = 1^2 + (-2)^2 + 4.6^2$, a little thought will show that the expression for $\|\mathbf{v}\|$ agrees with Pythagoras' Theorem. We say that **x** is a *unit vector* if $\|\mathbf{x}\| = 1$. It is possible to give a more geometric expression for the dot product in terms of the norm:

$$\mathbf{v} \cdot \mathbf{w} = \|\mathbf{v}\| \times \|\mathbf{w}\| \cos \theta,$$

where θ is the angle between the vectors.

Because $\cos 90^\circ = 0$, we say that nonzero vectors **v** and **w** are *orthogonal* ('at right angles') if $\mathbf{v} \cdot \mathbf{w} = 0$.

Above, we have expressed vectors in terms of a particular basis (coordinate system) namely $\mathbf{i} = (1, 0, 0)$; $\mathbf{j} = (0, 1, 0)$; $\mathbf{k} = (0, 0, 1)$. Take **v**, for example:

$$\mathbf{v} = (1, -2, 4.6) = 1\mathbf{i} - 2\mathbf{j} + 4.6\mathbf{k}$$

Taken together, the vectors **i**, **j**, **k** are called an *orthonormal basis* because every pair of them is orthogonal, and each one of them is a unit vector. Such an expression for a vector **v** in terms of the basis is called a *linear combination*; and the numbers 1, -2, 4.6 are called the *components* of the vector (in this basis). The *dimension* of a vector space is the number of vectors in any basis. In the above example there are three, namely **i**, **j** and **k**; so **v** lies in a 3-dimensional space. Equally, you could count the number of **v**'s coordinates.

A *linear operator* M takes each vector **v** in space V to another (or occasionally the same) vector, called M(**v**), which is also in V. By *linear* we mean that M has the properties:

$$M(\mathbf{v}) + M(\mathbf{w}) = M(\mathbf{v} + \mathbf{w})$$

$$M(a\mathbf{v}) = aM(\mathbf{v}) \qquad [\text{So } M(\mathbf{0}) = \mathbf{0}]$$

... for all **v** and **w** in V, and for all scalars a.

Included among the linear operators are all of the change-of-basis transformations mentioned above, namely rotations, reflections and shears, and stretching by a non-zero factor. All of these take a basis to another basis, and so are reversible, with the inverse denoted M^{-1}. But there are other linear transformations, say N, which in effect 'squish' vector space V, of say 5 dimensions, into a subspace W of V, of say 3 dimensions. This happens if N sends certain non-zero vectors to **0**. Such transformations N do not have an inverse. For example, let V be a two-dimensional space, with coordinates (x, y), and let N be the operator that always sends (x, y) to (x, 0), so N(3, 4) = (3, 0) for instance. Then N squishes all the points in the (x, y)-plane on to the x-axis. We cannot invert N because N also sends (3, 5) to (3, 0), so we cannot send vector (3, 0) back to where it came.

Another tricky feature of operators is that their effect usually depends on the order in which they are applied. Applying M first and then L is written LM; applying L first and then M is written ML. In general, LM ≠ ML.

To sum up, real vector spaces are 'the mathematics of arrows in n-dimensional space'. The dimension can be large of even infinite. Linear operators are 'linear transformations such as rotations, shears, etc., which act on vectors'.

More important for quantum theory are ***complex vector spaces*** whose scalars are complex numbers rather than real numbers. Whenever a complex vector is expressed in components, each component is a complex number. The rules remain somewhat similar. Suppose $\mathbf{v} = (v_1, v_2) = (2+3i, 7.5-4i)$ is a vector in a 2-dimensional complex vector space. Here $v_1 = 2+3i$ is the first component of $\mathbf{v}$; and $v_2 = 7.5-4i$ is the second component. Similarly let $\mathbf{w} = (w_1, w_2) = (7+0i, 2-1i)$.

As well as the vector space rules, we need a complex analogue of the dot product. The notation for this is '$\langle \mathbf{v} | \mathbf{w} \rangle$' and it is now always called the *inner product*. In general, it is a complex number. Calculating $\langle \mathbf{v} | \mathbf{w} \rangle$ is similar to calculating the dot product – it is obtained by multiplying corresponding components and then adding – ***except*** that we first take the complex conjugate of each component of the vector on the left of the inner product:

$$\langle \mathbf{v} | \mathbf{w} \rangle = v_1{}^*w_1 + v_2{}^*w_2$$

$$= (2-3i)7 + (7.5+4i)(2-i)$$

$$= [14-21i] + [19+0.5i]$$

$$= 33 - 20.5i$$

It can be shown that the inner product has the following properties. Let $\mathbf{u}$, $\mathbf{v}$, $\mathbf{w}$ be any vectors and c be any scalar, then:

$\langle \mathbf{u} | \mathbf{v}+\mathbf{w} \rangle = \langle \mathbf{u} | \mathbf{v} \rangle + \langle \mathbf{u} | \mathbf{w} \rangle$ also: $\langle \mathbf{u}+\mathbf{v} | \mathbf{w} \rangle = \langle \mathbf{u} | \mathbf{w} \rangle + \langle \mathbf{v} | \mathbf{w} \rangle$

$\langle \mathbf{u} | c\mathbf{v} \rangle = c\langle \mathbf{u} | \mathbf{v} \rangle$ also: $\langle c\mathbf{u} | \mathbf{v} \rangle = c^*\langle \mathbf{u} | \mathbf{v} \rangle$

(In the latter case note that the conjugate c* appears outside rather than c.)

$$\langle \mathbf{u} | \mathbf{v} \rangle = \langle \mathbf{v} | \mathbf{u} \rangle^*$$

When both sides of the inner product are identical, the result is always a real number:

$$\langle \mathbf{0} | \mathbf{0} \rangle = 0; \qquad \text{and for } \mathbf{u} \neq 0,\ \langle \mathbf{u} | \mathbf{u} \rangle > 0$$

Our example was in a 2-D space. In a 5-D space, the general formula would be:

$$\langle \mathbf{v} | \mathbf{w} \rangle = v_1{}^* w_1 + v_2{}^* w_2 + v_3{}^* w_3 + v_4{}^* w_4 + v_5{}^* w_5$$

In an infinite dimensional space, the sum continues indefinitely.

Most calculations essentially go through as before. The *norm* ('length') of a vector **v** is given in the new notation:

$$\| \mathbf{v} \| = \sqrt{\langle \mathbf{v} | \mathbf{v} \rangle} \qquad \text{(It is still a real number.)}$$

A complex vector space, of any dimensionality, which has an inner product, is called a ***Hilbert space***, and this is the fundamental mathematical object in the quantum recipe. (There are some additional technicalities, but these are not of interest here.)

At last, we approach the actualities of the world. Figure 4.6 shows the configuration of a couple of balls on a snooker table. The white ball is at the position with coordinates (x_W, y_W), and the green is in position (x_G, y_G). We can summarise this arrangement by regarding it as a single point, (x_W, y_W, x_G, y_G), in a 4-D vector space, which is called the ***configuration space*** of this system. Had there been 7 balls, then the configuration space would have had 14 dimensions.

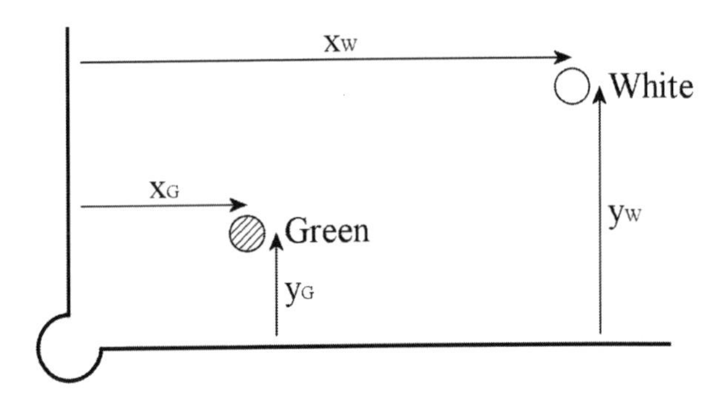

Figure 4.6: Configuration of the white and green balls on a snooker table

The number of *degrees of freedom* of a physical system is the dimensionality of its configuration space, as in the above examples. Next, suppose that in Figure 4.6 the two balls are connected by a rigid rod, then the configuration is completely specified by (x_W, y_W, θ_G), where θ_G is the direction of G, as viewed from W: Now the system has 3 degrees of freedom. Alternative coordinates may be used to describe configurations, but they will always have the same degrees of freedom; for example, (x_G, y_G, θ_W).

The quantum recipe: part II, physics

Quantum mechanics investigates systems (most typically, but not necessarily, extremely small) that obey quantum rules. We'll call such systems *quantum systems*.

The wavefunction, ψ

Suppose we have a quantum system with k degrees of freedom, then the *wavefunction*, often denoted ψ (the Greek letter psi), is a function that associates a complex number c with each point in the configuration space: $\psi(x_1, x_2, \ldots, x_k) = c$. If we write $\mathbf{x} = (x_1, x_2, \ldots, x_k)$, we can shorten this to $\psi(\mathbf{x}) = c$. Strictly speaking,

ψ is also a function of time, and whenever this is relevant, it is written as an additional argument: $\psi(\mathbf{x}, t) = c$. When t is omitted, it is implicit that the discussion of the wavefunction pertains to some fixed time, say t_0.

Let p be a proton and e an electron, and suppose for the moment that they can roam anywhere in ordinary 3-D space. We can write a typical point in the configuration space of this system as $(x_p, y_p, z_p, x_e, y_e, z_e)$, where x_p is the x-coordinate of the proton and so on. Then, at this point, $\psi(x_p, y_p, z_p, x_e, y_e, z_e) = c$. We still abbreviate this to $\psi(\mathbf{x}) = c$, but now the arguments are more informative than a bland list of labels.

The notation $\mathbf{x} = (x_p, y_p, z_p, x_e, y_e, z_e)$ tells us that the second item in the list "is associated in some way with" the y component of the proton's position. The scare quotes are a warning that, in standard quantum mechanics, protons and electrons do not have well-defined positions as such. Each merely manifests a position property in the context of a position measurement, as will become clearer later.

An important fact about the wavefunction, ψ, is that it can be regarded as a vector in an infinite-dimensional Hilbert space, some of whose properties have been sketched above. When so regarded, it is sometimes written $|\psi\rangle$, and called a *ket-vector*, or *ket* for short. The wavefunction of the entire universe is sometimes written Ψ, or $|\Psi\rangle$.

The Hilbert space (call it H) is infinite dimensional because, for each possible $\mathbf{x}$, the functions δ_x (which take extremely large values very close to $\mathbf{x}$, and are zero elsewhere) form a basis. Since there are an infinite number of points $\mathbf{x}$ in configuration space, there are an infinite number of δ_x in the basis of H. So H is infinite dimensional. Such a δ_x is called a *Dirac delta function (centred on* $\mathbf{x}$*)*, named after its inventor. These functions are acknowledged to be, strictly speaking, mathematically invalid; crucially, this does not affect the physics, however, provided these functions are used with care. In what follows I will be assuming that you know a little about integration. Some of the most important facts are:

- Integration is about finding the area under the graph of a function f
- In regions where the function becomes negative, the integral is negative
- Integration is linear: $\int (af + bg)\, d\mathbf{x} = a\int f\, d\mathbf{x} + b\int g\, d\mathbf{x}$
- Integration is in some sense the converse of differentiation

The square of the norm of ψ can be expressed as an integral:

$$\|\psi\|^2 = \langle\psi|\psi\rangle = \int \psi^*(\mathbf{x})\, \psi(\mathbf{x})\, d\mathbf{x}$$

Strictly, this is the integral over the entirety of configuration space. In order for ψ to be a physically valid wavefunction, $\|\psi\|^2$ must be finite and non-zero. Given any valid ψ, we can divide it by its norm to ensure that the above integral is 1; such a wavefunction is then said to be *normalised*. It is usual to do this. How are we to understand (normalised) ψ in terms of physics? The answer is that, given any region R of configuration space, the probability that, when measured, the configuration of the system lies within R is:

$$\text{Prob(system found to be within R)} = \int_R \psi^*(\mathbf{x})\, \psi(\mathbf{x})\, d\mathbf{x}$$

In short, ψ indirectly gives us a way to calculate where we are more likely or less likely to find particles once we have performed an experiment. This is the simplest and most important instance of the *Born Rule*, named after Max Born, who discovered it in 1926, and later won a Nobel Prize in consequence.

If φ (phi) is another wavefunction, their inner product can also be expressed as an integral:

$$\langle \psi | \varphi \rangle = \int \psi^*(\mathbf{x})\, \varphi(\mathbf{x})\, d\mathbf{x}$$

Here are some miscellaneous ***facts about wavefunctions***: If the system under consideration is a single particle, then ψ can be regarded as being a complex-valued function in ordinary 3-D space. Otherwise, we have to remember that it is a function taking values in configuration space. The function ***0***, which is zero everywhere, is not a valid wavefunction because it cannot be normalised. It is a safe and standard practice to normalise a wavefunction, *except in cases where we want to consider it as either encountering, or as part of, another quantum system*. Multiplying wavefunction ψ by a constant change of phase makes no physical difference, *with the same exceptions*.

If ψ and φ are valid wavefunctions, then their sum, $\psi + \varphi$, is always a valid wavefunction (with the trivial exception where this sum is ***0***). The sum is called the *superposition* of ψ and φ.

General measurements

This section is quite technical, and much of it is not essential. It is included here for completeness. One fact that will be used later is that the linear operator associated with measuring momentum is '$-i\hbar\partial/\partial x$'. Knowing this, you may wish to skip down to the next subsection, ***Planck's Constant***.

So far, we have been discussing position measurements, but it is possible to give rules for each type of measurement (say a measurement of momentum, spin, or energy, etc.). Physicist John Bell was a very strong advocate of restricting measurements to position measurements. His reasoning was that, although different types of measurements require different types of apparatus, the results are always displayed in terms of position: a pointer points to the left or right; a spot at a particular position on a fluorescent screen brightens; pencil marks are made in certain locations in a scientist's notebook; etc. There is no necessity to include other types of measurements in the theory, and to add them risks introducing inconsistencies.

Nevertheless, physicists have found it convenient to add such procedures for general types of measurements, and here I will merely sketch how this is done. We have a given physical situation with its known configuration, **x**. The initial wavefunction ψ is also known because of the way in which the system has been prepared. Associated with each type of measurement there is a specific linear operator M. Once given M (which we can look up in a quantum recipe book), we can find certain wavefunctions, ε, such that:

$$M(\varepsilon) = r\varepsilon$$

That is to say, the effect of applying M to ε just multiplies ε by the real number r. Each such ε is called an *eigenvector* (of M), and each such r is called an *eigenvalue* (of M). There are many eigenvectors, ε_i, for operator M, each with its own eigenvalue, r_i. In fact, there are enough such eigenvectors to form a basis of the configuration space. We can therefore write ψ in terms of them:

$$\psi = c_1\varepsilon_1 + c_2\varepsilon_2 + c_3\varepsilon_3 + \ldots$$

Here the c_i are complex numbers; and each eigenvector ε_i is a wavefunction in its own right.

The (general) ***Born Rule*** is that, when we make a measurement (of the type associated with operator M), the outcome that we will observe is one of the eigenvalues of M, say r_a, with probability $|c_a|^2$ (= $c_a{}^*c_a$). To confirm this, we can perform the experiment repeatedly to see that the predicted probability distribution builds up. If our initial wavefunction is already an eigenvector of M, say $\psi = \varepsilon_a$, it follows that we are guaranteed to observe the value r_a with probability 1.

Suppose now that we have a single-particle system that begins in some arbitrary state, ψ, and that when we measure

M for this particle the result is r_a. If we soon afterwards repeat this measurement on the same particle, then we will find that the experimental result is invariably (i.e., with probability 1) r_a. It would seem that the state of the system has changed from ψ to ε_a, and moreover that this physical change was somehow brought about by the process of measurement. This is what has come to be known as the *collapse of the wavefunction*.

The ***position operator***: Let us follow the above rituals for the operator M that is associated with measuring the position of a single particle. For simplicity, I'll assume that the particle is constrained to move along the x-axis only, and that its spin is 0. For any given position a, the Dirac delta function centred on a is an eigenvector for M with eigenvalue a:

$$M(\delta_a(x)) = a\,\delta_a(x)$$

There are infinitely many equations here – one for each position a. We can expand ψ in terms of this basis (because the basis is infinite, the sum becomes an integral):

$$\psi(x) = \int \psi(a)\,\delta_a(x)\,da$$

Here $\psi(a)$ is analogous to a typical coefficient c_a in the expression: $\psi = c_1\varepsilon_1 + c_2\varepsilon_2 + c_3\varepsilon_3 + \ldots$, and likewise δ_a corresponds to the eigenvector ε_a. So $\psi(a)$ can be used to calculate probabilities. By the general ***Born Rule*** above, the probability that the measurement outcome is close to position a is proportional to $|\psi(a)|^2$. To be more specific, let R be the interval between p and q, then:

$$\text{Prob}(a \text{ in } R) = \int_R \psi(a)^*\psi(a)\,da$$

Or, changing the label from a to x (which is no more than a

typographical difference):

$$\text{Prob}(x \text{ in } R) = \int_R \psi(x)^*\psi(x)\, dx$$

We have recovered the simple Born Rule by going through the ritual of the general ***Born Rule*** above. It turns out that the linear operator M associated with measuring position x is simply 'Multiply by x'. Using similar methods, it can be shown that the linear operator associated with measuring momentum is '$-i\hbar\partial/\partial x$'.

As remarked above, associated with every possible type of measurement there is a linear operator M. There is little physical motivation for this association, except that it is such that the rules will produce the correct results. For instance, when making a position measurement, the same operator is used whether one is employing a ruler or a radar echo to measure distances. So, M gives little insight as to what is going on physically. (A certain restriction is placed on M so as to ensure that all of its eigenvalues are real rather than complex; but this mathematical technicality is not relevant to us here.)

Planck's Constant

Scientists only became aware of the quantum nature of the world at the beginning of the twentieth century. Why wasn't this seen before? The first insight into this strange world came in 1901, when Max Planck, in order to explain a certain experimental anomaly, proposed a model in which energy was not continuous, but came in minuscule discrete lumps ('quanta'). He proposed *Planck's Law*: that the energy E of a particle of light (a photon) of frequency f was given by:

$$E = hf$$

Here h is *Planck's constant,* whose value is astonishingly minuscule – about $h = 6.6 \times 10^{-34}$ Joule-seconds. For example, visible light has a frequency f of about 10^{15} cycles per second. So, the energy of a photon of visible light is roughly 10^{-18} Joules. For a long time, Planck believed that his constant was no more than a mathematical ploy which happened to give the correct answer to his particular problem. But other scientists soon saw Planck's constant cropping up in a multitude of situations, and quickly came to realise that it was a fundamental constant of nature. It had not been discovered earlier because of its tiny size.

The symbol ħ ('h-bar') is often used as shorthand in quantum calculations: By definition $\hbar = h / 2\pi$.

Schrödinger equation

In all of the above we have ignored time for the most part. But if we prepare a quantum system to be in an initial state with wavefunction ψ, how does this state develop over the course of time? The answer is given by the *Schrödinger equation*:

$$i\hbar\ (\partial/\partial t)\ |\psi(\mathbf{x}, t)\rangle = H\ |\psi(\mathbf{x}, t)\rangle$$

Here $|\psi(\mathbf{x}, t)\rangle$ is the wavefunction defined over a configuration space with k degrees of freedom; the extra argument t has been included, which of course denotes time. $(\partial/\partial t)$ is a mathematical operator (from calculus), which tells how the wavefunction changes over time. Both i and ħ you know.

The only item remaining is H, which is a linear operator related to the energy of the system under consideration. Without it, the Schrödinger equation is devoid of any physical meaning. How are we to calculate it? There is a chicken and egg paradox here. If the system were a system of classical particles, then the energy of the system at any instant would depend not only on the positions $(x_1, x_2, \ldots, x_k)$ of the particles, but also on their momenta (p_1, p_2, ..., p_k). The relevant space for calculating the classical energy is called *phase space*, a typical point of which is (x_1, x_2, ..., x_k, p_1, p_2, ..., p_k).

There is a problem here. Take the ith position and momentum, x_i and p_i, for example. It is a very well attested experimental fact that the combined error, Δ, in measuring these properties cannot be reduced below ħ/2. In symbols: $\Delta x_i \times \Delta p_i > \hbar/2$. Werner Heisenberg discovered this in 1927, and for this reason it is called the *Heisenberg uncertainty principle*. He initially attributed this to a disturbance of the system caused by making the measurement. Nowadays the physical cause is more commonly attributed to the wave nature of ψ. In this chapter, however, we are studiously avoiding physics, and are instead concentrating on the mathematics of the quantum recipe. It is sufficient for our purposes to know that the above properties can never, under any circumstances, be simultaneously and exactly determined.

The problem can now be stated: How can we ascertain the energy operator H for a given quantum system? The solution is a 'magical' four-step process:

- Pretend that we have a classical system of particles

At any given time, t, these have known positions, momenta, masses, charges, etc. The forces acting between them will be fully known, and so the energy will be known.

- Calculate the classical energy E of the system for each point in phase space

This will be a known function f of the point in phase space. (Function f will also depend upon some known constants, for example, the masses and charges of the particles: these will also appear in the formula for f.) In symbols:

$E = f(x_1, x_2, \ldots, x_k, p_1, p_2, \ldots, p_k)$

- Replace each x_j in f with the operator 'multiply by x_j'

Some authors symbolically decorate these operators with hats (^), but I won't do this here: the change is to be understood from the context.

- Similarly, replace each p_j in f with the operator '$-i\hbar\partial/\partial x_j$'

Here you really do have to make the substitution. Thus transformed, formula f is the required linear operator H. We are done! This process is slightly ambiguous: the product xp is identical to the product px, but the corresponding operators do not commute (this is another feature of the Heisenberg uncertainty principle). In this situation, one just chooses the solution that works. Particles with spin have to be treated with their own methods, as we shall see.

Classical energy

I'm going to say just a little about this, so that the energy operator is made more explicit. The simplest possible example is a classical particle constrained to move in a single direction, x, and we will also assume that it is not moving close to the speed of light. Its kinetic energy is $\frac{1}{2}mv^2$; and its potential energy, V, depends on the external forces (if any) acting on the system – it is a function of position. The momentum of the particle is given by p = mv, and so its kinetic energy, $\frac{1}{2}mv^2$, can be rewritten as $p^2/2m$.

Energy E = kinetic energy + potential energy:

$$E(x) = p^2/2m + V(x)$$

The corresponding energy operator is:

$$H(x) = (1/2m)(-i\hbar\partial/\partial x)^2 + V(x)$$

$$= -(\hbar^2/2m)(\partial/\partial x)^2 + V(x)$$

If the particle is allowed to roam in three dimensions, we can replace x by the 3-D vector $\mathbf{x} = (x, y, z)$. We can also replace $(\partial/\partial x)^2$ by its 3-D equivalent:

$$\nabla^2 = (\partial/\partial x)^2 + (\partial/\partial y)^2 + (\partial/\partial z)^2$$

We arrive at:

$$H(\mathbf{x}) = -(\hbar^2/2m)\nabla^2 + V(\mathbf{x})$$

In situations where no external forces are acting, V is 0 everywhere and can be ignored. In such cases the particle is said to be *free*.

Energy is a conserved quantity. So, the only situation in which a physical system's energy changes over time is when energy is moved in or out of it. This is modelled by V being a function of time as well as position, but we will not be considering such examples.

About this mathematical notation

You may already be familiar with the notion that if f is a function of x, then df/dx is the function (of x) that gives the slope of f at each point x. For example, you probably know that if $f(x) = 3x^2$, then $df/dx = 6x$. This can also be written $(d/dx)(3x^2) = 6x$. The slope function df/dx is called the *derivative* of f. Differentiation is linear, meaning that $(d/dx)(af + bg) = a(d/dx)f + b(d/dx)g$, where a and b are any numbers. We can take the derivative twice, to find the 'slope of the slope' of f:

$$(d/dx)^2(3x^2) = (d/dx)(6x) = 6$$

The traditional notation for $(d/dx)^2$ is d^2/dx^2, and it is called the *second derivative.*

When we have a function of **several** variables, say $g(x, y, z) = x^2 \cdot \sin(y) + z^3 x$, then we can still differentiate it with respect to x, by regarding y and z as constants. In doing this it is customary to make a change of notation, replacing d with ∂:

$$(\partial/\partial x)g = (\partial/\partial x)\ (x^2 \cdot \sin(y) + z^3 x) = 2x \cdot \sin(y) + z^3$$

In the first term, sin(y) is treated as a constant; and in the second, z^3 is so treated. The function $(\partial/\partial x)$ is called the *partial derivative* with respect to x. We can apply $(\partial/\partial x)$ again to get the *second partial derivative,* which turns out to be $(\partial/\partial x)^2 g = 2 \cdot \sin(y)$. As an exercise, find the second partial derivatives of g with respect to y and z. Adding these parts together, you will find that:

$$\nabla^2 g = (\partial/\partial x)^2 g + (\partial/\partial y)^2 g + (\partial/\partial z)^2 g$$

$$= 2\sin(y) - x^2 \cdot \sin(y) + 6xz$$

Time evolution of the wavefunction

What happens to the wavefunction ψ as it evolves over time according the Schrödinger equation? Here are a couple of precise facts: First, if ψ is initially normalised at $t = 0$, then the Schrödinger equation keeps it normalised at all times. Second, the Schrödinger equation is linear. If ψ, φ are wavefunctions, and c, d are complex numbers, then, for all time t

$$(c\psi + d\varphi)(\mathbf{x}, t) = c\psi(\mathbf{x}, t) + d\varphi(\mathbf{x}, t)$$

That is to say, the Schrödinger equation conserves superpositions over time.

Other facts about the evolution of the wavefunction will be given in merely qualitative terms. Let us take the simplest case of a single particle which is constrained to move in direction x only, so ψ is a function of x and t alone. Moreover, we will assume that the particle is free: $V(x) = 0$ everywhere.

Our first examples concern a particle whose initial position is quite well known. Let us suppose that, at time 0, the particle lies within a small interval $a < x < b$. So, $|\psi(x, 0)| > 0$ only if $a < x < b$, and is zero elsewhere. When calculated, Schrödinger's equation shows that, as time passes, ψ will spread out in space more and more. The smaller the initial interval (a, b) is, the faster the spreading will be. If the interval is extremely small, the wavefunction will approximate to the Dirac delta function, and its initial spreading will be explosive. Such a wavefunction, initially lying within interval (a, b) is called a *wave packet*. At one point Schrödinger believed that particles were in reality wave packets, but he abandoned this idea when he realised that these packets spread out indefinitely according to his equation.

At the other extreme is a wavefunction whose position is completely unknown, but whose momentum is known exactly. In this case, the Schrödinger equation leaves the wavefunction unaltered (apart from a change of phase, which is only detectable under specific circumstances, previously mentioned). To be more specific, at time 0, the wavefunction has the form

$$\psi(x, 0) = Ae^{iBx}$$

And at time t, the Schrödinger equation has evolved it to

$$\psi(x, t) = Ae^{iBx} \cdot e^{-Ct}$$

Here, e^{-Ct} is the change of phase; A, B, and C are certain constants, not specified here. A is a constant over the entirety of space, reflecting our complete ignorance as to where the particle is. This is not physically realistic, and this is reflected in the maths because ψ is not normalisable. It is better to consider a wave packet in which the amplitude A varies with x, and declines so that it is zero outside the interval $a < x < b$. When this interval (a, b) is very large, it approximates to the situation where the momentum is known exactly. (More will be said about such momentum states in the explanation of the Young's Slits experiment.)

If the particle is now allowed to move anywhere in space, then the above qualitative considerations apply to each direction (x, y, z) independently. When we are dealing with a single particle, the wavefunction, for any given time, can be regarded as existing in ordinary 3-D space. The same applies to a beam of particles, each of which passes through any relevant apparatus *independently* – that is to say, without influencing one another.

Suppose wavefunction ψ is normalised. It must be the case that, if we evaluate ψ at positions increasingly distant from the origin, then $|\psi|$ must tend to 0 very rapidly. This leads to the concept of the *support* of the wavefunction ψ, at given time t_0. Select a positive real number w. Let R be the region of those points $\mathbf{x}$ for which $|\psi(\mathbf{x}, t_0)| > w$. It is possible to calculate the probability of finding the particle within this region; moreover, we can adjust w so that this probability has any value we wish, for example 0.99. We say that R is the *support* of ψ (at the 99% confidence level, at time t_0). In practice, we do not make specific calculations, but we often know with great confidence that the wavefunction is negligible outside some region R, which we call the support of ψ (at time t_0).

The support R can replace the interval (a, b) in the above qualitative discussions. The same arguments go through, but apply to a wider class of wavefunctions (those which are not strictly zero outside the interval). In particular, we can allow for more physically-realistic wave packets if we define them to be wavefunctions whose support is a small region of space (at some initial time). In addition, the concept of support is used in textbooks to visualise the *orbitals* of electrons in atoms (here w is typically chosen so as to give a 90% probability of finding the electron within the boundary of the orbital).

To sum up: For free particles, the effect of Schrödinger's equation is to disperse the initial wavefunction ψ ever more widely. The more the wavefunction is constrained in any direction, the more rapidly it will disperse. The wavefunction will approach, but never quite attain, a state in which position is wholly unknown and momentum is exactly known. This can be monitored crudely by showing how the support of the wavefunction spreads over time.

This qualitative discussion is by no means complete. In situations where there is more than one particle, we must remember that ψ is not a function of physical space, but of configuration space, which has a huge dimensionality. (The wavefunction is also a function of time, of course.) Moreover, this section has not considered the effect of external potentials V(x) on the time evolution of the wavefunction. Animations exist on the Internet; search for "wavefunction animation" on YouTube, for instance.

The quantum recipe: part III the recipe itself

Here is a summary of how the quantum recipe is used in practice:

1. *Divide the world into two parts: the quantum system under investigation and the laboratory*
2. *Prepare the quantum system to have a certain, known wavefunction*
3. *Model the quantum system as it passes through the apparatus using the Schrödinger equation*
4. *Use the Born Rule to predict the probability of each possible outcome at each measuring device*
5. *Repeat the experiment many times to check whether these statistical predictions are confirmed*

Quantum recipe is somewhat vague – for example it is unclear as to where the division between the quantum system and the laboratory should be made. In this chapter we are only concerned with the quantum recipe as a predictive tool; and it is superbly successful in this role. In contrast, what (if anything) does quantum mechanics tell us about the nature of what is actually going on in the world? This is a highly controversial question, which will be the topic of following chapters.

Meanwhile, the final task is to explain the experimental findings discussed earlier in terms of the quantum recipe. This will be done only qualitatively and informally. You can be extremely confident that physicists have proven these results in full mathematical detail.

Explaining Young's Slits experiments

The Young's Slits experiments are sketched in Figure 4.1. We know the momentum p of particles from the source, but we have little information about their position. We also know that they are travelling in the x-direction. Then the wavefunction ψ has the general form (see Griffiths & Schroeter, 2018, p. 55 for a more formal account):

$$\psi(x, y, z, t) = Ae^{i(Bx-Ct)}$$

The amplitude A of the wavefunction is almost constant, reflecting our comparative lack of knowledge of the particle's position (in a more careful analysis, A is a gently sloping function of y and z). B and C are constants related to the momentum p.

In the initial discussion of the Young's Slits experiments, it was mentioned that scientists in the nineteenth century explained interference using the concept of waves. In quantum mechanics we have the wavefunction, but the analogy is far from exact because water waves have heights that are real numbers, whereas wavefunctions are complex functions. Nonetheless we will make the convention that phase 0 (i.e., Bx = Ct) of the wavefunction corresponds somewhat to the crest of the water wave. The wavefunction ψ then corresponds to the train of water waves sketched here:

At any given time, t, the crests of the waves are vertical lines (lines of constant x); and these waves propagate from left to right, as shown.

The portion of the beam coming from the source, and which passes through the barrier in Figure 4.1, is very narrow; the rest of the beam can be ignored. Therefore, this beam can essentially be regarded as having the form just described.

Single slit experiment

The single slit experiment is almost as shown in Figure 4.1, except that only one slit, say the Upper, is open. Within the narrow slit, the support of ψ_U in the y-direction is restricted to the slit's width. The result is that, as described earlier in the subsection ***Time evolution of the wavefunction***, the wave ψ_U exiting the barrier disperses rapidly in the ±y-direction. The phase changes of ψ_U are analogous to semi-circular ripples emerging from the location of the slit (their degree of spreading has been exaggerated). The beam has widened by the time it reaches the target, and the pattern of intensity is shown in Figure 4.2(a). The same Figure also shows what the pattern would have been if the Lower slit had been open instead.

Double slits experiment

When both slits are open, two beams ψ_U, ψ_L emerge from the barrier, from the Upper and Lower slits respectively. The right half of Figure 4.1 shows this very schematically: There are two sets of semi-circular ripples, each centred on its own slit, and the lines show where the phases of each partial beam are 0.

Here, the combined wavefunction ψ is the superposition of ψ_U and ψ_L, with equal weights:

$$\psi = \frac{1}{\sqrt{2}} |\psi_U\rangle + \frac{1}{\sqrt{2}} |\psi_L\rangle$$

The disanalogy between wavefunctions and water waves now becomes apparent. Instead of the **height** of the combined water wave at each location, we must compute the **squared amplitude** of the combined wavefunction, $|\psi|^2 = \psi^* \psi$.

The wavefunctions ψ_U, ψ_L each have roughly equal, constant amplitudes, say A. However, at any given location their phases will usually differ. As time increases, the phases of ψ_U and ψ_L will each decrease at a common rate. Consequently, the **phase difference** remains constant over time, and this is what is important physically. In places where the phases happen to be equal, $|\psi|^2$ is 2A. In places where the phases point in diametrically opposite directions, $|\psi|^2$ is 0. In places where the phases differ by some other amount, $|\psi|^2$ will have some intermediate value. The resulting combined intensity $|\psi|^2$ on the target may be shown to be as depicted in Figure 4.2(b).

In both of the above experiments, the intensity of the source can be turned down sufficiently that only one particle is within the apparatus at any time. When this happens, the pattern of intensity on the target builds up as a sequence of localised flashes. This characteristic behaviour is described (but not explained) by the Born Rule, given earlier.

'Monitored' double slits experiment

Such specks suggest that what is being observed has at least some of the characteristics of a particle. However, any attempt to monitor which slit the supposed particle went through, by any apparatus whatsoever, has the effect of destroying the interference pattern of Figure 4.2(b). What is observed instead is an intensity pattern that is the sum of the two curves in Figure 4.2(a); this is a smooth curve with a single peak.

A problem with discussing this experiment is that measuring devices are highly complex, consisting of many particles. Tim Maudlin has devised an idealised measuring device whose active part is a single proton. It is convenient to set up a coordinate system for Figure 4.1. The origin of the coordinates is at the very centre of the barrier, between the slits. The x-axis points directly to the right, in the plane of the page. The y-axis is also in the plane of the page, but points up. The z-axis points out of the page, towards you, the reader. Because the slits in the barrier are parallel to the z-axis, we can ignore the z-coordinate in most calculations.

We imagine that a tiny chamber is made, which is completely enclosed within the barrier. It is located at the origin of the coordinate system, between the slits. This chamber is minuscule in the x and z directions, but long enough in the y direction that it **almost** reaches each slit: Suppose the slits are 2.2d apart, as measured between their inner edges (d is very small), then the chamber extends from y = +d to y = -d in the y-direction. The proton is trapped within this sealed chamber, and, at the start of each run of the experiment, is centred within it.

In an experimental run, a single electron is released from the source; it passes through the slits, and causes a flash on the target. Because of its opposite charge, the electron attracts the proton as it passes through the slits. By the end of the run, if the electron passed through the Upper slit, the proton will be found near the upper end of the chamber, with position coordinates approximately (0, 0, +d). Similarly, if the electron passed through the Lower slit, the proton will end up near (0, 0, -d).

We consider the configuration space for this situation. It is six-dimensional, having coordinates $(x_e, y_e, z_e, x_p, y_p, z_p)$.

However, some of these dimensions are unimportant: z_e can be ignored because the whole setup of Figure 4.1 is largely uniform in the z-direction. The proton is constrained to move in the y-direction. As a result, the configuration space is effectively three-dimensional: (x_e, y_e, y_p), and we can picture this if we draw y_p as the third coordinate.

After the electron has passed through the barrier by a sufficient distance (say $x_e > 1$) to have completed its interaction with the proton, the combined wavefunction will be a superposition

$$\psi = \frac{1}{\sqrt{2}} |\psi_U\rangle\, |\text{proton up}\rangle + \frac{1}{\sqrt{2}} |\psi_L\rangle\, |\text{proton down}\rangle$$

This combined wavefunction now consists of two disjoint branches, as we shall see. Proton up corresponds to $y_p = +d$; and proton down to $y_p = -d$. The partial (for $x_e > 1$) support of the first term of ψ is the region in configuration space for which

$$(x_e > 1,\ y_e \text{ any value},\ y_p = +d)$$

Similarly, the partial (for $x_e > 1$) support of the second term of ψ is the region in configuration space for which

$$(x_e > 1,\ y_e \text{ any value},\ y_p = -d)$$

It is easy to see that these regions are disjoint partial slices within configuration space. (Make a rough sketch of these regions, using (x, y, z) coordinates. You will only be able to sketch a portion of these regions, and indicate where they extend to infinity.)

Because of this separation, there can be no interference between branches. The Upper branch produces the intensity graph to the left of Figure 4.2(a), and the Lower branch produces the graph to the right. These sum to produce the overall intensity plot, already mentioned, which shows no interference fringes.

Tim Maudlin's measuring device is highly impracticable, but it does have the great advantage that the configuration space may be readily visualised. He comments:

> This separation of the wavefunction in configuration space has nothing to do with any observer of visible result: the separation occurs via strict Schrödinger evolution due to the strong coupling of the electron and proton.
> (Maudlin, 2019, pp. 57-58)

He also gives very informative diagrams of this experimental situation (pp. 53, 57).

Explaining basic spin experiments

Spinors

Before explaining experiments involving spin, we have to discuss spinors, which are the way in which spin is represented mathematically in quantum mechanics. Spin is usually dealt with separately from the wavefunction, as follows. Set up an (x, y, z) coordinate system in the laboratory, as in Figure 4.3 for example. Choose an arbitrary direction in space, say z.

$|z\uparrow\rangle$ is a symbol expressing the fact that the particle is in the state z-spin up.

$|z\downarrow\rangle$ is a symbol expressing the fact that the particle is in the state z-spin down.

These symbols are vectors in a two-dimensional complex vector space called *spinor space*; and in fact they form a basis in this space, called the *z-basis*. It is important to notice that these two symbols $|z\uparrow\rangle$, $|z\downarrow\rangle$ are orthogonal as spinors; even though the directions $z\uparrow$, $z\downarrow$ are at 180° in physical space. So here the enclosing ket-brackets '$|\ \rangle$' are essential to avoid confusion.

Spin in any spatial direction can be expressed in terms of a complex linear combination of these basis vectors:

$c|z\uparrow\rangle + d|z\downarrow\rangle$ where c and d are complex numbers

Spinors are invariably normalised so that $|c|^2 + |d|^2 = 1$. If c and d are both real, then this simplifies to $c^2 + d^2 = 1$.

Let us use the x spatial direction instead. It turns out that we can express x-basis spinors $|x\uparrow\rangle$ and $|x\downarrow\rangle$ in terms of the z-basis spinors

$$|x\uparrow\rangle = \frac{1}{\sqrt{2}}(|z\uparrow\rangle + |z\downarrow\rangle)$$ (here both c and d are $1/\sqrt{2}$)

$$|x\downarrow\rangle = \frac{1}{\sqrt{2}}(|z\uparrow\rangle - |z\downarrow\rangle)$$ (here $c = 1/\sqrt{2}$, but d is $-1/\sqrt{2}$)

Suppose $C|x\uparrow\rangle + D|x\downarrow\rangle$ is an arbitrary spinor expressed in the x-basis, where C and D are complex numbers with $|C|^2 + |D|^2 = 1$. Let us express this in the z-basis.

$$C|x\uparrow\rangle + D|x\downarrow\rangle = \frac{C}{\sqrt{2}}(|z\uparrow\rangle + |z\downarrow\rangle) + \frac{D}{\sqrt{2}}(|z\uparrow\rangle - |z\downarrow\rangle)$$

$$= \frac{C+D}{\sqrt{2}}|z\uparrow\rangle + \frac{C-D}{\sqrt{2}}|z\downarrow\rangle$$

$$= c\,|z\uparrow\rangle + d\,|z\downarrow\rangle$$

The last line above defines c and d for our example. So, to go from the x-basis to the z-basis, given C and D, we take $c = (C+D)/\sqrt{2}$, and $d = (C-D)/\sqrt{2}$.

Conversely, suppose we wish to go the other way – from the z-basis to the x-basis. This assumes that we know c and d and we must find expressions for C and D from them. You can check that the solutions are: $C = (c+d)/\sqrt{2}$, and $D = (c-d)/\sqrt{2}$. That is to say, we just swap upper- and lowercase letters in the pair of equations just discovered. You can also check that, if c and d are normalised, such that $|c|^2 + |d|^2 = 1$, then so are C and D – and vice versa.

In ordinary space, consider the unit vector r which lies in the x-z plane, and is such that it is tipped from the vertical z by an angle θ towards x. We want to write spinors in terms of the r-basis $|r\uparrow\rangle$, $|r\downarrow\rangle$. First of all, as usual, we write each of these in terms of the z-basis. It turns out that:

$$|r\uparrow\rangle = \cos(\theta/2)\ |z\uparrow\rangle + \sin(\theta/2)\ |z\downarrow\rangle$$

So, in this situation $c = \cos(\theta/2)$ and $d = \sin(\theta/2)$. These coefficients are already normalised by Pythagoras' Theorem (expressed in the form $\sin^2 + \cos^2 = 1$). Similarly, it turns out that:

$$|r\downarrow\rangle = \sin(\theta/2)\ |z\uparrow\rangle - \cos(\theta/2)\ |z\downarrow\rangle$$

So here $c = \sin(\theta/2)$ and $d = -\cos(\theta/2)$, which are also normalised.

For a slightly more completeness, it turns out that the spinors corresponding to the y spatial direction are:

$$|y\uparrow\rangle = \frac{1}{\sqrt{2}}\ |z\uparrow\rangle + \frac{i}{\sqrt{2}}\ |z\downarrow\rangle \qquad \text{(here } c = 1/\sqrt{2}\text{, and } d = i/\sqrt{2}\text{)}$$

$$|y\downarrow\rangle = \frac{1}{\sqrt{2}}\ |z\uparrow\rangle - \frac{i}{\sqrt{2}}\ |z\downarrow\rangle \qquad \text{(here } c = 1/\sqrt{2}\text{, and } d = -i/\sqrt{2}\text{)}$$

In all the experiments discussed in this book, the particles are travelling in a beam. So – in order to avoid having complex coefficients – it is convenient to choose coordinates such that the beam is travelling in the y-direction.

A word of caution: we haven't really said anything about

physics here. Spinors are elegant mathematical tools, invented in 1927 by Wolfgang Pauli, so as to give the correct answers to all possible experiments involving spin in any physical direction. They provide little insight as to what is going on in the world. In particular, the expressions for going from one basis to another are mathematical conventions devised by Pauli.

Spinors and the Stern-Gerlach apparatus

Consider Figure 4.3, where the Stern-Gerlach apparatus is oriented in the z-direction. Particles from the source have their spins oriented in random directions. That is to say, a particle chosen at random from the beam has spin:

$$c|z\uparrow\rangle + d|z\downarrow\rangle$$

Where c and d are complex numbers, normalised so that $|c|^2 + |d|^2 = 1$.

A new and important fact about spinors is that the probability that such a particle is deflected up by the z-oriented S-G apparatus is $|c|^2$, and the probability that it is deflected down is $|d|^2$.

The above spinor corresponds to a certain spin direction in physical space, call it **n**, which can be expressed in (x, y, z) coordinates: $\mathbf{n} = (n_x, n_y, n_z)$, normalised so that $n_x^2 + n_y^2 + n_z^2 = 1$. Given the (real) components n_x, n_y, n_z of the direction, it is possible to derive the (complex) components c, d of the spinor – and vice versa; but I'm not giving these formulae here.

Had the randomly chosen particle been spinning in the opposite spatial direction, **-n**, then it can be shown that its spinor would have had the general form:

$$Ad|z\uparrow\rangle - Ac|z\downarrow\rangle$$

The most important fact here is that c and d have swapped position.

Less significant – because they can be ignored when calculating probabilities – are that A is some complex number with $|A| = 1$, and the plus sign has become minus. For a spinor corresponding to spatial direction **–n**, the probability of being deflected up is $|d|^2$, and the probability of being deflected down is $|c|^2$.

To sum up, choose a particle at random coming from the source. It is spinning in a certain direction, and the probability of it being deflected up or down by the apparatus are $|c|^2$ and $|d|^2$ respectively. Had this particle been spinning in the opposite direction, then these probabilities would have been reversed.

Particles coming from the source are spinning randomly in any direction. Thus, it is equally likely that a particle chosen at random from the beam would be spinning in either of the opposing directions mentioned above. The particles spinning about spatial direction **n** are thus counterbalanced by those spinning in direction **-n**. The conclusion is that, overall, half of the particles in the beam are deflected up, and half are deflected down. This shows that spinor theory correctly predicts that the particles land on the target as depicted in Figure 4.3. In this discussion, we have been concerned with the **actual** spin-state of each particle coming from the source, **prior to measuring** it, i.e., prior to it hitting the target.

First spin experiment

In this experiment, recall that a beam of z-spin up particles is prepared by replacing the target with a barrier having a slot so that only particles measured as being z-spin up can go through. There is a tricky point here: suppose that a particle comes from the source with initial spin:

$$\frac{1}{\sqrt{10}}|z\uparrow\rangle + \frac{3}{\sqrt{10}}|z\downarrow\rangle$$

Then this particle can still go through the slit, with probability

1/10 – as do many other particles with different initial spins. How does it come about that, when a beam of such prepared particles are again measured for their z-spin, all are found to be z-spin up? The answer is that the preparation device (the S-G apparatus with slotted target) still constitutes a measuring device – even for those particles that pass through the slot. The wavefunction collapses to the 'observed' value implicit in the position of the slot. In particular, the new spin state for **every** particle that successfully emerges from the slot is:

$$1|z\uparrow\rangle + 0|z\downarrow\rangle = |z\uparrow\rangle$$

Thus, when this prepared beam is measured for a second time, all particles are found to be z-spin up, with probability $|c|^2 = |1|^2 = 1$. So the theory explains the results of the first experiment.

Second spin experiment

This experiment was to prepare a beam of z-spin up particles, and measure them in the x-direction. The previous section showed, that if we have particles with spin

$$c|z\uparrow\rangle + d|z\downarrow\rangle$$

Then if we measure their x-spin, we will get

$$C|x\uparrow\rangle + D|x\downarrow\rangle \qquad \text{where } C = (c+d)/\sqrt{2}; \text{ and } D = (c-d)/\sqrt{2}$$

As we have seen, because of the collapse postulate, preparing the beam to be z-spin up means that, for all particles in the prepared beam, $c = 1$ and $d = 0$. It follows that both C and D are $1/\sqrt{2}$. So, the theory predicts that half the particles are x-spin up, and half are x-spin down. This is in agreement with the results observed in this experiment.

Fourth spin experiment

We will skip the third experiment, and explain the final one, in which we prepare a beam of z-spin up particles, measure the x-spin, select those which are x-spin up, and finally measure the z-spin again. This involves a source and three S-G apparatuses: S-G(1) is oriented in the z-direction, with a slot to prepare the z-spin up particles; S-G(2) is oriented in the x-direction, with a slot to prepare the x-spin up particles; S-G(3) is oriented in the z-direction for the final measurement – no slot is needed.

Here, the first two S-G apparatuses are like those of the experiment described previously. Now though, we make a slot in the second S-G apparatus, and allow the x-spin up particles pass through. Because of the collapse of the wavefunction, the spin state of the beam emerging from S-G(2) is entirely:

$$|x\uparrow\rangle$$

What do we get when we measure this beam with S-G(3), which is oriented in the z-direction? From the previous section,

$$C|x\uparrow\rangle + D|x\downarrow\rangle = c|z\uparrow\rangle + d|z\downarrow\rangle$$

where $c = (C+D)/\sqrt{2}$, and $d = (C-D)/\sqrt{2}$

Putting C = 1 and D = 0 into this formula, we get

$$|x\uparrow\rangle = \frac{1}{\sqrt{2}}|z\uparrow\rangle + \frac{1}{\sqrt{2}}|z\downarrow\rangle$$

And we have seemingly lost all information about z-spin. The calculations go through in a similar way if we instead choose down-spins in the appropriate direction for either or both of the apparatuses S-G(1) or S-G(2). This completes the theoretical explanation of the observed results of the fourth experiment.

Third spin experiment

This experiment consists of a pair of Stern-Gerlach apparatuses. The first apparatus, S-G(1), is used to prepare a beam of r-spin up particles. It is therefore rotated by an angle θ in the x-z plane, from z towards x. Its slotted target is similarly rotated, so as to let only $|r\uparrow\rangle$ particles through. The second apparatus, S-G(2), is oriented in the z-direction. Its target (without a slot) is used to measure the probability that the particles emerging from S-G(2) are z-spin-up or z-spin-down. In actuality, this probability is measured by comparing the intensities at the two regions shown on the target.

The setup is depicted in Figure 4.3, when this is understood as follows: the particle source is regarded as a "black box" in which S-G(1) is hidden; the prepared beam emerging from this source is $|r\uparrow\rangle$; and S-G(2) is fully shown. The experimental findings are that the probability of agreement (either up-up, or down-down) is $\cos^2(\theta/2)$.

The spinor for the prepared beam, $|r\uparrow\rangle$, was given in the previous section:

$$|r\uparrow\rangle = \cos(\theta/2)\ |z\uparrow\rangle + \sin(\theta/2)\ |z\downarrow\rangle$$

So here $c = \cos(\theta/2)$ and $d = \sin(\theta/2)$. The probability that particles exiting S-G(2) will be found in state $|z\uparrow\rangle$ when they hit the target is predicted to be $c^2 = \cos^2(\theta/2)$, according to quantum theory. This is what is found experimentally, and can be confirmed for all angles θ. The experiment is thus confirmed by the theory in the case of r-up and z-up.

For completeness we must confirm the result for the probability of agreement in the **down** direction. To do this, we move the slot on the target of S-G(1), so as to prepare a beam of $|r\downarrow\rangle$ particles. Now the appropriate spinor is:

$$|r\downarrow\rangle = \sin(\theta/2)\ |z\uparrow\rangle - \cos(\theta/2)\ |z\downarrow\rangle$$

The probability that this result is also down, when a measurement is made in the z-direction, is $d^2 = (-\cos(\theta/2))^2 = \cos^2(\theta/2)$, as before.

Physical meaning of spinors

In many situations, spin is unimportant: For a single particle without spin, or for a particle in an experiment where spin is irrelevant, or for a beam of such particles, the Schrödinger equation has arguments x, y, z, t: $\psi(x, y, z, t)$. When spin is important, then the configuration of the particle is supplemented by its spinor, $|s\rangle$, to give $\psi(x, y, z, |s\rangle, t)$. Here, $|s\rangle$ is an additional feature of the configuration of the particle. It is traditional to express $|s\rangle$ in the z-basis. In this case the wavefunction can be expressed as a superposition:

$$\psi(x, y, z, |s\rangle, t) = c\ \psi(x, y, z, |z\uparrow\rangle, t) + d\ \psi(x, y, z, |z\downarrow\rangle, t)$$

Where c and d are normalised complex numbers, so that $|c|^2 + |d|^2 = 1$. Abbreviate this to

$$\psi(x, y, z, |s\rangle, t) = c\ \psi_{\uparrow}(x, y, z, t) + d\ \psi_{\downarrow}(x, y, z, t)$$

In principle, the wavefunctions $\psi_{\uparrow}$, $\psi_{\downarrow}$ could be computed for the given physical situation. In practice, a general description is often sufficient. For example, suppose we have a z-oriented Stern-Gerlach apparatus. The non-uniform magnetic field between its poles is such that a $|z\uparrow\rangle$ particle in the beam has wavefunction $\psi_{\uparrow}$, and is invariably deflected up by the apparatus. Conversely, this same magnetic field is such that a $|z\downarrow\rangle$ particle has wavefunction $\psi_{\downarrow}$, and is invariably deflected down.

A particle spinning in an **arbitrary** direction $|s\rangle$ has a wavefunction given by the superposition above. It is still invariably deflected either up or down, but with probabilities given by the general Born Rule: Prob(spin-up) = $|c|^2$; and

Prob(spin-down) = $|d|^2$. As previously mentioned, even particles spinning very close to $|z\uparrow\rangle$ can very occasionally be deflected down (and vice versa). In practice we can often ignore the wavefunction, and deal with spin in isolation, but the above discussion explains how these two things are related. We replace the wavefunction by very general descriptions, such as, "The apparatus causes the particles to be deflected either up or down..."

How does spin affect the wavefunction in physical terms? Let $\mathbf{x} = (x, y, z)$ be a position in space. Suppose there is a non-uniform magnetic field $\boldsymbol{B} = \boldsymbol{B}(\mathbf{x})$; this will be the case for example between the poles of the magnet in a S-G apparatus. Recall from the very beginning of this section that the physical direction of a particle's spin, **n**, can be deduced from c and d. The potential V of the particle that appears in the wavefunction is given by the following vector dot product (see Maudlin, 2019, p. 62):

$$V(\mathbf{x}) = -\frac{e}{m}\left(\frac{\hbar\,\mathbf{n}}{2}\right) \cdot \boldsymbol{B}(\mathbf{x})$$

Where e is the charge, and m is the mass of the particle. When **n** is replaced by **-n**, this has the effect of changing the sign of V. To obtain the expression for $\psi_\uparrow$, we would set $\mathbf{n} = (0, 0, 1)$ to give $V(\mathbf{x})$. To obtain the expression for $\psi_\downarrow$, we would instead use $-V(\mathbf{x})$.

(A complication not mentioned in the description of **all** of the above experiments is that charged particles travel in circles in magnetic fields. Additional apparatus is used to counter this unwanted effect.)

Explaining interferometer experiments

First interferometer experiment

The first experiment with the interferometer of Figure 4.4 is to measure x-spin at position **A**, using an x-oriented S-G apparatus. When this is done, it is found that all particles at **A** are x-spin up. The mathematical formalism predicts this correctly. In the previous section we found that

$$C|x\uparrow\rangle + D|x\downarrow\rangle = c|z\uparrow\rangle + d|z\downarrow\rangle$$

Where $c = (C+D)/\sqrt{2}$, and $d = (C-D)/\sqrt{2}$. The input beam at the bottom-left has been prepared $|x\uparrow\rangle$, which means that $C=1$, and $D = 0$. It follows that $c = d = 1/\sqrt{2}$: the z-down and z-up beams are equal in intensity, and follow widely separated routes through the interferometer. When these are recombined into one at the top-right, the beam has the form $c|z\uparrow\rangle + d|z\downarrow\rangle$, i.e., the RHS of the above expression. We saw in the previous section that to express this in terms of the x-basis, we must calculate $C = (c+d)/\sqrt{2} = 1$, and $D = (c-d)/\sqrt{2} = 0$. So, the theory predicts that the recombined beam is $|x\uparrow\rangle$, in line with experiment.

In passing, it was mentioned that each sub-beam **considered in isolation** appeared to contain no information about x-spin, which can be confirmed by placing x-oriented S-G apparatuses at **B** and **C**. Doing these checks essentially repeats the ***Fourth spin experiment*** above.

If the source of the interferometer experiment had been prepared x-spin down, then the output at **A** would also be x-spin down. Spinor theory confirms this, as you may check for yourself by repeating the above calculations, but starting with $C = 0$ and $D = 1$.

Second interferometer experiment

This second experiment involves a mysterious "do nothing box". This box allows a beam to pass through it – apparently unaltered, because it has no observable effect when used in any of the basic spin experiments, described and explained earlier. This box contains magnets causing the physical direction **n** associated with the spinor to turn through 360° (2π radians), leaving **n** unchanged. The underlying spinor, however, has its phase changed by half of this: the phase is $e^{i\pi} = -1$

$$c|z\uparrow\rangle + d|z\downarrow\rangle \text{ has become } -c|z\uparrow\rangle + -d|z\downarrow\rangle$$

Note that for an undivided beam, the probabilities of observing spin-up or spin-down are unchanged: $|c|^2 = |-c|^2$; and $|d|^2 = |-d|^2$. So, the basic spin experiments are unaltered by the box.

Let us analyse what happens in the interferometer experiment when the "do nothing box" is placed in the lower beam at **C**. After passing through the box, this half-beam, with spinor $(1/\sqrt{2})\ |z\downarrow\rangle$, has become $(-1/\sqrt{2})\ |z\downarrow\rangle$. The upper half-beam remains, as before, $(1/\sqrt{2})\ |z\uparrow\rangle$. When the beams are recombined, they become:

$$\frac{1}{\sqrt{2}}|z\uparrow\rangle - \frac{1}{\sqrt{2}}|z\downarrow\rangle$$

And this, by definition, is the spinor $|x\downarrow\rangle$. The "do nothing box" has converted an x-up input beam into an x-down output beam. Even though each half-beam, investigated in isolation, seems to contain no information about x-spin, they must – taken together – contain this information.

You can make further checks showing that it doesn't matter whether the "do nothing box" is placed at **B** or **C**; or on the initial orientation of the input beam. In all such cases, the output beam will have the opposite orientation to the input beam.

More about the physical meaning of spinors

A spinor $|\mathbf{s}\rangle$ is an abstract mathematical entity. This one object can be represented in different bases, as we have seen, for example:

$$|\mathbf{s}\rangle = c|z\uparrow\rangle + d|z\downarrow\rangle = C|x\uparrow\rangle + D|x\downarrow\rangle$$
$$\text{where } |c|^2 + |d|^2 = 1 \text{ and } |C|^2 + |D|^2 = 1$$

In any basis, we may add a complex phase, $e^{i\omega} = \cos(\omega) + i \sin(\omega)$, without altering $|\mathbf{s}\rangle$; for example, $|\mathbf{s}\rangle$ could equally well be written:

$$|\mathbf{s}\rangle = e^{i\omega}c|z\uparrow\rangle + e^{i\omega}d|z\downarrow\rangle$$

Phases of spinors can usually be ignored, except in cases where the spinor represents only a portion of a wavefunction – as in the ***Second interferometer experiment*** above. Thus, phase for spinors plays the same role as does phase for wavefunctions.

For a spin-½ particle, to convert from abstract mathematical units to physical units, we multiply spin vectors by $\hbar/2$. This was seen earlier in the expression for the potential V of a particle in a Stern-Gerlach apparatus.

Further reading

In addition to Maudlin's excellent *Philosophy of Physics: Quantum Theory* (2019), on which much of this chapter is based: Griffiths & Schroeter's *Introduction to Quantum Mechanics* (2018) is a readable, modern undergraduate text.

Chapter 5

Copenhagen and entanglement

By the late 1920s it was widely believed that quantum mechanics had reached its final and perfected form, at least in non-relativistic situations. There was universal agreement as to how quantum mechanical experiments were to be set up, performed, and the results predicted from the mathematical formalism: all this was described in the previous chapter, under the term "the quantum recipe". Despite this consensus, the **physical meaning** of the formalism and its application was recognised to be obscure: what, if anything, was the quantum recipe telling us about the character of our universe – particularly about its goings-on at the subatomic level? The most influential method of trying to understand this was based upon a philosophical position, now popularly called the "Copenhagen interpretation". This position was so dominant that it and the science of quantum mechanics came to be regarded as being one and the same. Any scientist, no matter how distinguished, who questioned the Copenhagen interpretation, was accused of closing their mind to the inescapable scientific facts. They were disparaged and had their arguments ignored or misrepresented; some lost their careers. Even Einstein was treated with condescension by the majority of quantum physicists.

Then, in 1935, Einstein, together with colleagues Podolsky and Rosen (EPR), published a paper that made a *prima facie* convincing case that quantum mechanics must be an incomplete theory – not giving a full account of what was going on in the world – unless a certain spooky process was possible. This was that measurements made by scientists in one part of the universe could, under certain circumstances, instantly **change** the state of a physical system at a remote location. The second location, perhaps spacelike separated from the measurements, should (one would think) be insulated from them by reason of distance. This proposed property of the universe, called *non-locality*, seemed to be the height of unreason. But, unless it was true, quantum mechanics had to be incomplete.

Eminent scientists responded to the EPR paper, but for many years discussions remained at the level of philosophical debate about the character of physical reality – akin to those between Plato and his followers. Then, in 1964, John Bell formulated a mathematical inequality that could lead to an experimental test which would decide these matters. By 1982 Alain Aspect and his team had succeeded in performing experiments that thoroughly tested Bell's inequality. The experimental findings are subtle and will be discussed fully later. But one conclusion is certain – our universe is indeed non-local. The Copenhagen School put a simplistic gloss on all this: "Einstein was wrong and quantum mechanics is vindicated." This misrepresentation continues to this day – even in the teaching of some of our most prestigious universities. The truth is that, before EPR considered (and then rejected) the possibility that the universe might be non-local, no scientist had seriously considered this issue. Inspired by Einstein, Schrödinger (1935a) was the first to recognise non-locality within the theory, and to give a general mathematical analysis of it. He was appalled and named the quantum mechanical process that led to this property: *entanglement*.

Adam Becker (2018) documents the above history in terms of both the science, and of the human failings that some eminent scientists exhibited. This chapter will give the history of these events, focussing principally on the science, but mentioning issues of personality where necessary. I have gone back to the original papers of the scientists concerned; in one case I have used an official (refereed) translation from German into English.

The Copenhagen interpretation

The "Copenhagen interpretation" is the name now given to a system that **extends the quantum recipe**, by appending to it a number of philosophical assertions. It was developed by Niels Bohr, Werner Heisenberg, and others, so they might provide an answer to the question: what can science tell us about the character of what is going on in the quantum world? In crude terms, the Copenhagen answer is 'nothing' or 'almost nothing', but a more refined and accurate response will be presented below.

The Copenhagen interpretation was influenced by the ideas of the eighteenth-century German philosopher Immanuel Kant. Its more recent (1920s) antecedents were the logical positivists such as Otto Neurath and Rudolf Carnap, whose philosophy was based on the *verification principle*: "No statement is meaningful unless it can be tested by experiment." Logical positivism denied any factual meaning to entire fields of knowledge, including ethics, aesthetics, and philosophy (except for its own position). It also imposed substantial constraints on what could legitimately appear in scientific theories. The fact that the verification principle is itself meaningless (according to its own criterion!) led to the downfall of positivism by the late 1960s.

Because Niels Bohr is regarded as the leader and foremost champion of the Copenhagen interpretation, the main focus in this chapter will be on his position. Bohr influenced the careers and philosophical mind-sets of generations of physicists. He won the Nobel Prize in 1922 for his early work giving a provisional theory of the hydrogen atom. As quantum theory developed, he performed a key positive role in preventing physicists falling into error by making erroneous assumptions. (Copenhagen is not a unified school of thought – its proponents sometimes had mutually-inconsistent understandings; but their conceptions all had the tenor of positivism.)

Here are some tenets of the Copenhagen interpretation:

- The validity of quantum mechanics is to be determined solely in terms of its success in predicting the outcomes of experiments.

This is an instance of the verification principle.

- The Copenhagen interpretation is the only possible way of understanding quantum mechanics.

To introduce anything other than predictive success into a physical theory is to introduce philosophy, which is forbidden by positivism. Following on from the above two points, but worth noting separately:

- Quantum mechanics and the Copenhagen interpretation are one and the same.

A common mistake of proponents of Copenhagen is to presume that their opponents must accept this as a necessary truth. But, if one rejects positivism, one is free to reject this assertion.

- Our only understanding of the quantum system under investigation **must be** rooted in everyday descriptions of what is going on in the laboratory.

Niels Bohr repeatedly stressed that: (a) natural language **must be** used in describing the experimental setup; (b) the apparatus **must be** described in terms of classical physics. This gave rise to his notion of complementarity, which will be discussed in the next section.

It is true that a scientist performing a routine experiment, to test the success of quantum mechanics as a predictive tool, indeed uses natural language to describe the setup. But a scientist can also use language creatively when trying to understand the world (just as Einstein modified the notion of time, for example). Moreover, although it might be the case that our best description of a measuring device can at best be given in wholly classical terms, this is hardly plausible as the final, objective physical truth. After all, by its nature, such a measuring device interacts with a quantum system in a specific and reliable manner. Mara Beller points out (1999, pp. 173-75) that Bohr incorrectly turns "may be" to "must be."

- Quantum mechanics gives a complete description of the world – one which is incapable of improvement.

Max Born and Werner Heisenberg made this audacious claim at the fifth Solvay Conference held in Brussels in 1927. They said that quantum mechanics was "a closed theory, whose fundamental physical and mathematical assumptions are no longer susceptible of any modification" (Becker, 2018, p. 46). Their confidence flies in the face of the founding principles of science, in which all theories are susceptible to radical change as new observational and experimental evidence arrives.

- 'Irrealism': Quantum mechanics gives us limited information about a quantum system, but only in the context of it being measured by us. Outside of this context, it is impossible – even in principle – to say anything about the reality of the quantum world.

It is uncontroversial that quantum measurements are highly contextual, and Bohr is rightly credited for emphasising this. The second sentence, however, was (and still is) disputed by Bohr's opponents: In the early decades of the twentieth century, there was no well-developed, realistic account of quantum goings-on in the absence of measurement. But surely it remains meaningful to affirm the reality of the universe – including its subatomic happenings – before the coming of life? And surely it is within the remit of scientists to attempt to investigate this?

Bohr's expression of the above tenet is unequivocal: "In our description of nature the purpose is not to disclose the real essence of the phenomena but only to track down as far as possible relations between the manifold aspects of our experience" (quoted in Stapp, 2004a, p. 64).

In practice, however, Bohr himself frequently took a softer line. He regarded a measurement as being completed, not when a human observer witnessed it (say, when a scientist examined a photographic plate), but when an irreversible change occurred in the apparatus (say, when a spot appeared on the plate). Bohr wrote:

> The description of atomic phenomena has in these respects a perfectly objective character, in the sense that no explicit reference is made to any individual observer ... As regards all such points, the observation problem of quantum physics in no way differs from the classical physics approach.
> (Quoted in Stapp, 2004a, p. 65)

(Bell discusses this ambiguity in detail: 1990, pp. 34-35.) This relaxed definition of 'measurement' allows one some limited grip on the reality of the unobserved world but it has two problems. First, the registering of a spot on a photographic plate does not constitute a measurement in positivist terms. Second, it undermines the claim that quantum mechanics is complete: the unobserved, uncontrolled, irreversible change (called by Bohr the "quantum of action") must be physically real – this is clear from its description – but does not feature in the quantum recipe, which only gives predictions as to what the scientists will observe. (Consider a photograph that is locked away and never looked at.)

- The stochastic nature of quantum phenomena is unavoidable.

Bohr wrote, "The fact that in one and the same well-defined experimental arrangement we generally obtain recordings of different individual processes makes indispensable the recourse to a statistical account of quantum phenomena" (quoted in Stapp, 2004a, p. 66).

It is universally accepted that present-day experiments can at best give us statistical knowledge about the world. Moreover, it is highly implausible that any advance, either in theory or experiment, could ever achieve more. Nonetheless, even if such uncertainty were an absolute limit to our knowledge, the possibility remains open that nature itself is deterministic. The Copenhagen School denies this latter possibility.

- Physical properties that are hidden from our experience (so-called *hidden variables*) cannot possibly exist.

This is in line with positivism. John Stewart Bell argued forcefully against this:

> The 'microscopic' aspect of complementary variables [e.g., exact particle positions] is indeed hidden from us. But to admit things not visible to the gross creatures that we are is, in my opinion, to show a decent humility, and not just a lamentable addiction to metaphysics. In any case, the most hidden of all variables ... is the wavefunction, which manifests itself to us only by its influence on the complementary variables.
> (Bell, 1987, p. 2)

- Physical characteristics implicit in the quantum recipe must be accepted as they stand, without the possibility of amendment.

Some of these are objectionable. For example, the observed system is described by a wavefunction, whereas the apparatus is described in classical terms: thus, the physical descriptions of two interacting physical systems are utterly different and incompatible. This unpleasant feature of the theory leaves no clue as to how an apparatus could be composed of atoms.

Niels Bohr and 'complementarity'

How are we to understand basic entities of physics, such as electrons and photons? In a crude manner of speaking, I have been using the words "particles", "waves", and "beams", but have cautioned that these terms are not to be taken literally. Perhaps, when quantum mechanics is better understood, humankind can develop entirely new concepts that better account for these entities. Niels Bohr would have none of this. For him, it will only ever be possible to describe such things in terms of the concepts that are familiar to us from our everyday lives.

Consider the Young's Slits experiments. In some situations, such as when interference occurs, the quantum system exhibits wavelike characteristics. In other situations, as when flashes occur on the target, the system exhibits particle-like characteristics. When the intensity of the source is extremely low, both characteristics seem to be exhibited together. Bohr's so-called explanation for this was that photons and electrons exhibited *complementary* properties depending on the experimental *context*. This, however, is to use the word "complementary" in order to paper-over a void in our understanding.

The idea of contextuality is extremely important, and Bohr performed a valuable service in the early decades of quantum mechanics in pointing out that what physical properties could in principle be observed in any experimental situation critically depended on the measuring devices actually present. If such devices are not present, it is a matter that can be argued either way as to whether or not unmeasured physical properties could be said to exist. (Bohr contended that they could **Not**.) Moreover, adding measuring devices **changes** the experiment. The canonical example is the disappearance of interference fringes in the monitored Young's Slits experiment. If one restricts complementarity to the more limited meaning of contextuality, then the idea is a valuable one (Stapp, 2004a, p. 280).

Using the word "complementarity" outside of its restricted meaning of contextuality was distinctive to Bohr; and of dubious value. Max Jammer called it "an extraneous interpretive addition to [the formalism]" (quoted in Beller, 1999, p. 238). For example, Bohr insisted that it was impossible to progress beyond traditional language and concepts in describing the novelties of quantum experiments.

Tim Maudlin is very critical of the alleged necessity of describing a physical quantum system using familiar concepts:

> Why think that Aristotle, or any other philosopher or scientist, who never considered quantum theory, had developed the right conceptual categories for characterizing everything physically real? The quantum state is a novel feature of reality on any view, and there is nothing wrong with allowing it a novel category: *quantum state*. This is, of course, not an informative thing to say, but it does free us from the misguided desire to liken the quantum state to anything we are already familiar with.
> (2019, p. 89, original italics)

To fully understand the above, one must realise that Maudlin defines a quantum state as a type of **physical system**; it is distinct from any **abstract mathematical description** we might have of it (perhaps by a wavefunction – or otherwise).

The historical use of the term "complementary" refers to different but mutually supportive properties. It is ironic that

Bohr, who insisted on the traditional use of language, defined quantum complementarity in terms of mutually contradictory properties. Whilst contextuality explains such seeming contradictions nicely, the concept of complementarity does not; instead, it adds confusion. As John Bell puts it:

> There is very little I can say about 'complementarity'. But I wish to say one thing. It seems to me that Bohr used this word with the reverse of its usual meaning. Consider the elephant. From the front she is head, trunk, and two legs. From the back she is bottom, tail, and two legs. From the sides she is otherwise, and from top and bottom she is different again. These various views are complementary in the usual sense of the word. They supplement one another, they are consistent with one another, and they are all entailed by the unifying concept 'elephant'. ... [In contrast, Bohr] seems to insist rather that we must use in our analysis elements which *contradict* one another, which do not add up to, or derive from, a whole.
> (Bell, 1989, p. 363, original emphasis)

Bohr was a great enthusiast for complementarity, zealously applying it to many fields. In psychology there was a supposed complementarity between 'thoughts' (masculine) and 'feelings' (feminine). This, of course, is no more than Bohr's rationalization for the prejudices of his era, which have since been refuted. He also saw complementarity as a way to resolve the mind-body problem. This purely verbal solution has no credibility. Beller notes that Bohr's claims in these fields were initially tentative, but hardened into certainties over the years, without any additional evidence being produced (Beller, 1999, chapter 12).

The mythological Einstein

A mythological narrative has grown up around Einstein, and

this false view still influences not only popular accounts of the history of quantum mechanics but also the academic teaching of the subject. The myth goes something like this:

> *Einstein rejected quantum mechanics because of his belief that "God does not play dice." Einstein proposed a number of thought experiments in an attempt to prove quantum mechanics wrong, but Bohr gave satisfactory explanations as to where Einstein was mistaken. Bell developed a test to see whether Einstein or quantum mechanics was correct. When Alain Aspect and other scientists performed the experiments to make the test, quantum mechanics was vindicated, and Einstein was proven to be wrong: "God does play dice."*

For an example of the kinds of attitudes that are summarised in italics above, see the interview with Rudolf Peierls (Davies & Brown, 1986, pp. 70-82). A detailed refutation of this myth will be given over the remainder of this chapter, but in the meantime, here is a preliminary line-by-line dissection of it:

- ***Einstein rejected quantum mechanics because of his belief that "God does not play dice."***

Einstein fully accepted quantum mechanics as a highly successful scientific theory which makes correct statistical predictions. In fact, he made many major contributions to it. Theoretical physicist Thibault Damour provides a convenient list (Damour & Burniat, 2017, pp. 145-6): **(1)** In 1905, in the course of investigating a phenomenon called the photoelectric effect, Einstein discovered light quanta (photons). For this work he was later awarded the Nobel Prize. **(2)** In 1906 he showed that material oscillators are quantised, and used this fact to explain the anomalous properties of diamonds. **(3)** In 1909 he showed that light simultaneously possesses particle-like and wave-like characteristics. This, from an alleged denier of quantum mechanics, is worth repeating loudly: **he showed that light simultaneously possesses particle-like and wave-like characteristics**. In 1916 he made three crucial theoretical advances: **(4)** First, he explained the energy and momenta of photons emitted and absorbed by an atom in terms of transitions between energy levels with that atom; **(5)** Second, he discovered a new quantum process that later gave rise to the invention of lasers; **(6)** Third, in anticipation of the later work of Heisenberg and others, Einstein introduced chance into quantum theory by using matrices. This is also worth repeating: **Einstein introduced chance into quantum theory**. **(7)** In 1924 he introduced the quantum statistics of a gas and used this to discover a new physical phenomenon, now called a "Bose-Einstein condensation".

So, Einstein accepted quantum mechanics, and used it to great effect. What he objected to was a questionable metaphysical position, the Copenhagen interpretation. He believed – along with fellow opponents of Copenhagen, Schrödinger and Bell – that it remained the vocation of scientists to try to get to grips with the realities of the universe as it exists in itself.

In a letter to Max Born dated 1926, Einstein made a famous remark to the effect that "God does not play dice" (Born, 1971, p. 91). As Bell commented, "There is a widespread and erroneous conviction that for Einstein determinism was always **the** sacred principle. The quotability of [the 'dice' remark] has not helped in this respect" (Bell, 1981a, p. 46). I suspect that Einstein himself may have come to regret his saying.

While Einstein greatly preferred theories which were deterministic, he never held that the universe had, as an absolute necessity, to be so. In a 1954 letter to Born, Pauli tries to clear up Born's misconception on this matter:

Einstein does not consider the concept of 'determinism' to be

> as fundamental as it is frequently held to be (as he told me emphatically many times), and he denied energetically that he had ever put up a postulate such as [quoting Born]: "the sequence of such conditions must also be objective and real, that is, automatic, machine-like, deterministic". In the same way, he *disputes* that he uses as criterion for the admissibility of a theory the question: "Is it rigorously deterministic?" (Born, 1971, p. 221)

- ***Einstein proposed a number of thought experiments in an attempt to prove quantum mechanics wrong, but Bohr gave satisfactory explanations as to where Einstein was mistaken.***

Bohr often missed the point that Einstein was trying to make with his thought experiments. His reply to Einstein's most powerful argument (the EPR paper which will be discussed in detail later) is notorious for its obscurity.

- ***Bell developed a test to see whether Einstein or quantum mechanics was correct.***

Bell's test ("Bell's Inequality") is not a simple case of Einstein versus quantum mechanics. Rather it is about Einstein versus the Copenhagen interpretation; and it involves many subtle issues. These include: realism vs. irrealism; locality vs. non-locality; essential randomness vs. determinism; and completeness vs. hidden variables.

- ***When Alain Aspect and other scientists performed the experiments to make the test, quantum mechanics was vindicated, and Einstein was proven to be wrong: "God does play dice."***

As we shall see, based on the evidence of Aspect's experiments, it was only on the issue of non-locality that Einstein was proven to be wrong. Moreover, it was Einstein and his associates who first considered the astonishing and unprecedented possibility that the world might be non-local. It was thus Einstein's work that eventually led to the discovery of this bizarre characteristic of the universe, which should therefore be added to the list of Einstein's contributions to quantum mechanics.

With regard to the other issues, we shall see that David Bohm developed a valid (thus necessarily non-local) interpretation of quantum mechanics that is realistic, deterministic, and with hidden-variables. The first two of these latter three characteristics, Einstein preferred in any theory. He regarded hidden-variables as being necessary, in order to embed the quantum mechanics in a more complete, more comprehensible theory. We shall discuss Einstein's reaction to Bohmian mechanics later. In any event, because Bohm's theory produces identical results to the standard theory, we can be certain that Bell and Aspect have **not proven any of the following three assertions**: that the quantum world is inescapably random; or that it must lack reality; or that the quantum recipe gives a complete description of this world.

The next task is to give a historical account of these events.

The Einstein, Podolsky and Rosen paper

In 1935, Einstein together with two colleagues, Podolsky and Rosen, wrote a seminal paper entitled "Can Quantum-Mechanical Description of Physical Reality be Considered Complete?" As its title suggests, the focus was on the completeness of the theory, and in order to reach a conclusion about this they had to found their arguments on several assumptions:

Objective reality and the aim of physical theory

Against the followers of the Copenhagen interpretation, Einstein and his colleagues (EPR) affirmed the objective reality of the physical world, "which is independent of any theory," and distinguished it from "the physical concepts with which the theory operates. These concepts are intended to correspond with the objective reality, and by means of these concepts we picture this reality to ourselves" (EPR, 1935, p. 777).

The correctness of quantum mechanics

There are two criteria by which to judge the success of a physical theory. The first of these is "Is the theory correct?" The only way of judging this is by "the degree of agreement between the conclusions of the theory and human experience." Nonetheless, as EPR make clear in the next sentence, this is not a judgement about human experience, but about physical reality, "This experience, which alone enables us to make inferences about reality, in physics takes the form of experiment and measurement" (p. 777).

EPR explicitly state that their paper does not question the correctness of quantum mechanics as a physical theory. Moreover, throughout the article they assume that all of its mathematical procedures produce predictions in agreement with experiment. It is important to reiterate this because popular accounts – and even some academics who should know better – have accused EPR of assuming that (to use a slogan) "quantum mechanics is wrong".

The completeness of a physical theory

The second criterion by which to judge a physical theory is that of its completeness, and this is the focus of the EPR paper. Their definition of *completeness* is straightforward; it is that *"every element of the physical reality must have a counterpart in the physical theory"* (emphasis original, p. 777).

What is an element of physical reality? As EPR rightly say, this cannot be determined by philosophical considerations, but must be found by observation and experiment; that is to say, by coming to grips with reality. They therefore do not attempt to give a comprehensive definition of such elements, but give a sufficient condition where it is safe to conclude that such an element of reality has been found. Their famous criterion is:

> [**Reality:**] *If, without in any way disturbing a system, we can predict with certainty (i.e., with probability equal to unity) the value of a physical quantity, then there exists an element of physical reality corresponding to this physical quantity.*
> (Italics original, p. 777)

As Tim Maudlin argues at length (2014, pp. 6-7), this criterion must be true by virtue of the words used in it. An experiment results in some measured value X with certainty. Because of the natural meaning of the word 'undisturbed', the system is by supposition unaltered by the measurement. Therefore, there must have been some fact Y about the system – as it existed even before the measurement – that caused X to be reliably observed. This Y is the element of reality corresponding to X. As Maudlin further explains, it also follows straightforwardly that fact Y must remain true of the system even if no measurement is made.

So, the validity of EPR's criterion is unassailable. But, as Maudlin points out, there is the subtle question as to whether one is **correct** to apply it in the specific experimental situation discussed in the EPR paper.

The completeness of quantum mechanics?

The Copenhagen School contended that quantum mechanics was complete: the only physical facts which pertain to a quantum system are the observed outcomes of experiments. Their confidence was based on the multitude of successes that the quantum recipe had in predicting these outcomes. The intention of EPR's paper was to show that ψ had to be supplemented by elements of reality, which could not be observed directly, but which could be inferred indirectly from experiments carried out elsewhere.

The EPR locality postulate

A locality postulate is not stated very clearly but is largely implicit in the EPR paper. It is a crucial prerequisite to understanding the argument being made. According to Maudlin (2014, p. 8) this postulate is in essence:

> A physical theory is *EPR local* if and only if according to the theory procedures carried out in one region [say A] do not immediately disturb the state of systems in any sufficiently distant region [say B] in any significant way. [I.e., the disturbance is not enough to cause any change in the results of any experiment.]
>
> (I have added clarifications [thus], Peter E.)

Roughly speaking, if EPR locality holds, and if A and B are as described, then nothing done at A can disturb B and so affect the elements of reality at B.

As pointed out by Maudlin, up until the time of the EPR paper all theories of physics throughout the history of science were known to be EPR local. For instance, Newton's physics is EPR local because, even though it is true that waving my arms here has the instant effect of making the star Betelgeuse jiggle slightly, this effect is vanishingly small. Special and general relativity are EPR local theories (provided we do not import supplementary quantum ideas into them) because, if A and B are spacelike separated, they cannot influence one another.

The plausibility of EPR locality is even more obvious if one takes its negation:

> A physical theory is *EPR non-local* if and only if according to this theory some procedures carried out in region A

> immediately cause changes in the results of experiments carried out in arbitrarily distant region B.

Einstein and his co-authors came to the natural but mistaken conclusion that EPR non-locality would render science impossible: scientists working at B would have their experiments ruined by large disturbances originating from many distant regions of the universe. They therefore assumed that regions A and B would effectively be physically insulated from one another, provided these places were sufficiently far apart. It was precisely because **Einstein, Podolsky and Rosen believed quantum mechanics to be a correct physical theory** that they assumed it must satisfy the criterion of EPR locality. No scientist before the publication of EPR's paper, including expert quantum physicists, had believed otherwise.

The controversial nature of the assumption of EPR locality only gradually became evident with the benefit of hindsight, many decades later. This is perhaps why Einstein and his co-authors were less explicit in defining it than they might have been.

The argument of the EPR paper

Consider a single particle constrained to move with one degree of freedom. Suppose that the wavefunction ψ of the particle happens to be an eigenstate for momentum. In this case, when the momentum of the particle is measured then the result is certain to be a particular fixed value, say p_0; moreover, such a measurement does not disturb the particle. This means that, in this situation according to the **Reality** criterion, there is an element of reality corresponding to p_0. (Here the position q is completely unknown.)

Because the **Reality** criterion is a sufficient condition, it does not preclude the possibility that there might be an element of reality corresponding to position, q. It could be that this element exists, but it does not become manifest in the experimental situation.

Alternatively, suppose that a similarly constrained particle is instead in an eigenstate, φ, for position. The same argument shows that, in this different quantum state, there is an element of reality corresponding to the position q_0 of the particle. (Here the momentum p is completely unknown.)

By the Heisenberg uncertainty principle (which EPR accept) there is no quantum state in which both p and q are known with certainty by direct measurement. It follows that:

> **(I)** The statements (S1) "Quantum mechanics is complete" and (S2) "Position and momentum, as described in detail above, can simultaneously correspond to elements of reality" cannot both be true.

For suppose "Quantum mechanics is complete", then all elements of reality must be derivable from the wavefunction. But, as a mathematical fact (Heisenberg), there is no wavefunction from which one can derive definite values for both p and q. So, if (S1) is true then (S2) is false: The statements cannot both be true together.

Figure 5.1: The EPR thought experiment. Particles a and b interact at the source. They separate, going to distant locations, A and B. Someone at A measures either the position or momentum of particle a. What effect, if any, does this have on particle b at B?

The authors' key idea, shown in Figure 5.1, is that perhaps we can find additional elements of reality by means of **indirect** measurements. The second half of the EPR paper imagines such an experiment. Suppose we have a pair of particles which interact for a time and then go their separate ways to distant locations A and B. The wavefunction for the combined system is Ψ. We cannot calculate the state of either particle, but must discover this by making further measurements.

Some tricky maths: measuring momentum

There now follows some difficult mathematics which you may wish to skip over:

EPR assume, in their equation (9), that the combined wavefunction has a certain special form:

$$\Psi(x_1, x_2) = \int \exp(i/\hbar \times (x_1 - x_2 + x_0) \times p)\, dp \qquad (*)$$

Here x_0 is an arbitrary constant; and exp(F) is an alternative way of writing e^F (for any expression F).

Suppose we want to measure the momentum p of the particle at A. Before the measurement we can express Ψ in the following extremely sketchy form:

$$\Psi(x_1, x_2) = \text{Sum over p of lots of terms like } A_p(x_1) \times B_p(x_2)$$

For those in the know, the sum is really an integral:

$$\Psi(x_1, x_2) = \int A_p(x_1) \times B_p(x_2)\, dp$$

Each $A_p(x_1)$ is a particular momentum eigenstate of the particle at A. We now measure its momentum, getting an observed value p_0. The wavefunction collapses instantaneously, and all terms in the sum vanish, except the one involving the observed value p_0. The changed wavefunction is now:

$$\Psi_{\text{after measuring momentum}}(x_1, x_2) = A_{p0}(x_1) \times B_{p0}(x_2)$$

For the special wavefunction Ψ defined in (*), $B_{p0}(x_2)$ is a momentum eigenstate for the particle at B. (This is the case because EPR deliberately chose the physical situation (and thus Ψ) so as to have this property.) By the law of conservation of momentum, this eigenstate will have eigenvalue $-p_0$.

Mathematics ends.

Element of reality for momentum

By our assumption of EPR locality, what we did at A had not disturbed the physical situation at B. But now, after our efforts, we can be sure that when the person at B performs a measurement of the second particle's momentum, she will inevitably find the value $-p_0$. By the **Reality** criterion, this means there must be an element of reality at B corresponding to this momentum. Moreover, this element must exist even if we do not perform the measurement.

Similar tricky maths: now measuring position

We can repeat the mathematics, now measuring **position**, denoted q, at A instead. There are a few differences: we must express Ψ as a different sum; this time of position eigenstates, let's call them *A*. After measuring position and finding q_0 we get a different collapse. For comparison, the formula after collapse is (again roughly):

$$\Psi_{\text{after measuring position}}(x_1, x_2) = A_{q0}(x_1) \times B_{q0}(x_2)$$

It turns out (by the special properties of (*)) that $B_{q0}(x_2)$ is a position eigenstate for the particle at location B, with eigenvalue $+q_0$. Here I've made the convention that position eigenstates are written in italics.

Element of reality for position

By our assumption of EPR locality, what we did at location A had not disturbed the physical situation at B. But now, after our efforts, we can be sure that when the person at B performs a measurement of the second particle's position, she will inevitably find the value $+q_0$. By the **Reality** criterion, this means there must be an element of reality at B corresponding to this position. Moreover, this element must exist even if we do not perform the measurement.

Concluding the EPR paper

Einstein, Podolsky, and Rosen take stock:

> We see therefore that, as a consequence of two different measurements performed on the first system [at location A], the second system [at location B] may be left in states with two different wavefunctions [either $B_{p0}(x_2)$ or $B_{q0}(x_2)$]. On the other hand, since at the time of measurement the two systems no longer interact, no real **change** can take place in the second system in consequence of anything that may be done in the first system. This is, of course, merely a statement of what is meant by the absence of an interaction between the two systems. *Thus it is possible to assign two different wavefunctions... to the same reality.*
> (Italics original, bold emphasis added, p. 779)

EPR now give a proof by contradiction (p. 780). As Fine notes (1986, p. 33), this can be shortened: As just now described, using the standard methods of quantum mechanics EPR have shown that "Position and momentum can simultaneously correspond to elements of reality." But this is the second statement in assertion **(I)** above, which is of the form "statements (S1) and (S2) cannot both be true." Therefore statement (S1) is false. We conclude that quantum mechanics is incomplete.

The paper finishes by asserting that "No reasonable definition of reality could be expected to permit [EPR non-locality]"; and by expressing confidence that a new theory, one which will provide a complete description of physical reality, is possible (p. 780).

Comment on the EPR paper

It is important to note that the two wavefunctions which describe what is allegedly the same physical situation at location B could not be more different:

> $B_{p0}(x_2)$ is a state in which **a component of momentum is known exactly** and position is completely unknown
> $B_{q0}(x_2)$ is a state in which **momentum is completely unknown** and a component of position is known exactly

Consider just momentum. The two descriptions (in bold above) are completely different. The same contrast applies if we consider what the above statements say about position.

The Einstein, Podolsky, Rosen paper is of crucial importance. This is clear as it remains today among the ten most-cited articles from *Physical Review*. In the following sections the immediate responses of three great scientists – Niels Bohr, Erwin Schrödinger, and David Bohm – are presented.

Bohr's response to EPR

The publication of the EPR paper came as a blow to the Copenhagen School. If its conclusions were accepted it would effectively demolish the Copenhagen interpretation. Niels Bohr put aside all other work in order to formulate an urgent rebuttal. Usually Bohr worked slowly, but in this instance his reply was submitted a mere two months later; and it appeared – in the same publication and with exactly the same title as the EPR paper – in October that year (Bohr, 1935). So pressing did Bohr consider this matter that, about a week before his manuscript was complete, he submitted a brief letter to the journal *Nature* sketching his response.

Bohr's paper begins by summarising the EPR paper. He writes that EPR have shown that it is possible to *"**predict** the value of any given variable"* (p. 696, all emphasis mine) in the particle at location B based on measurements carried out at a distant location, A, in situations where particles had interacted and then separated. But mere prediction is not surprising. This is analogous to picking up one of a pair of gloves, finding it fits the right hand, and instantly knowing that the other is the left one. What EPR have argued is that *there has been an instantaneous **change** in the physical state of the particle* at B.

Bohr goes on to write, "Such an argumentation, however, would hardly be suited to affect the soundness of quantum mechanical description" (p. 696). This is not relevant because EPR do not question the soundness (i.e., the correctness) of quantum mechanics anywhere in their own paper; instead, they question and go on to deny its completeness. Bohr argues that the Heisenberg uncertainty principle must be true because the process of measurement inevitably disturbs a quantum system (pp. 696-7). He also claims – in line with Copenhagen philosophy – that quantum mechanics requires "a final renunciation of the classical ideal of causality and a radical revision of our attitude towards the problem of physical reality" (p. 697). Because EPR claim to have avoided the problem of disturbance by dividing their system into two portions that are insulated from one another by distance, Bohr is duty-bound to refute this by the end of his exposition.

After a long preliminary discussion of the experimental arrangements for a Young's Slits experiment (pp. 697-699), Bohr asserts that, "The last remarks apply equally well to the special problem treated by [EPR] ... which does not actually involve any greater intricacies than the simple examples discussed above" (p. 699).

Bohr is at last ready to give a description of a possible experimental setup, including apparatus, which could at least in principle realise the EPR thought experiment and obtain data from it (pp. 697-700). As he concedes (footnote, p. 698) the setup he proposes is wildly impracticable but, as he correctly says, this does not affect his theoretical argument. The emphasis by Bohr on the nuts-and-bolts of the apparatus is typical of his method in dealing with thought experiments devised by Einstein.

Figure 5.2 gives a plan view of Bohr's experimental setup. Because of the extreme delicacy of the measurements, all of the apparatus must be fixed – unless otherwise indicated – to a rigid base, which forms the spatial reference frame for all measurements (pp. 697, 700). The base, or laboratory bench, is indicated schematically by the rectangle surrounding the Figure. All measurements in this experiment are made in the x direction shown. That is to say, for any object measured, we are invariably seeking the x component of either its position or momentum.

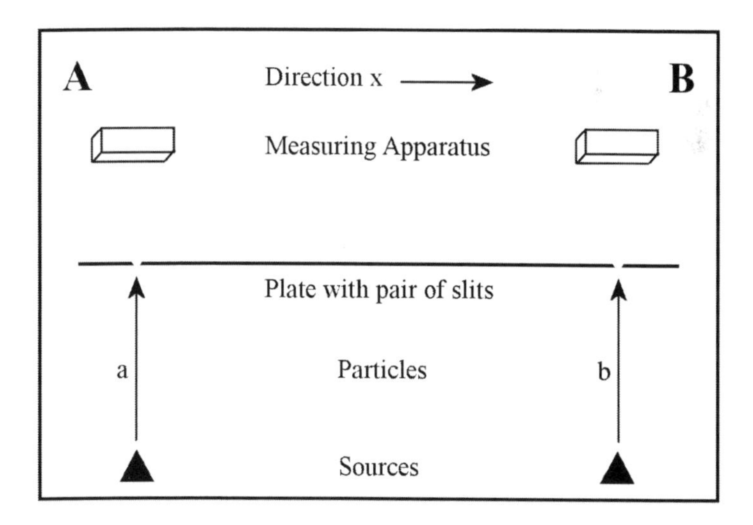

Figure 5.2: Plan view of Niels Bohr's experimental setup to implement the EPR thought experiment. All measurements are made in direction x.

Particles a and b are emitted from a pair of sources, and each of them passes through its own narrow slit. The slits are cut in a rigid, vertical plate, and they are separated by a large distance D. The momentum of the plate is measured both just before and just after the passage of the two particles. In order to do this, the plate cannot be fixed to the base, but must be free to move in the x direction. A piece of apparatus, not shown in the Figure but which is fixed to the base near one end of the plate, makes these measurements. From the latter, we can calculate the change in momentum ΔP of the plate. The conservation of momentum law gives us the sum of the momenta of the two particles just after passing through the slits: $p_a + p_b = -\Delta P$. But we can know nothing of the difference between these momenta at that instant.

The distance between the particles immediately after leaving the slits is well known because the narrow slits are widely separated (by distance D): $q_b - q_a = D$. But nothing is known about the sum of their positions in this direction. (By Heisenberg's uncertainty principle, we cannot know the plate's position because we have just measured its momentum; and $q_a + q_b$ is the position of the plate's midpoint.)

We can fix apparatus to the base in regions A and B in order to make measurements. Prior to carrying out the experiment we are free to choose the appropriate apparatus to measure either position or momentum in either location. This apparatus is shown schematically in the Figure as a pair of boxes.

Comments on Bohr's experimental setup

- The plate entangles the particles
- The preparation of the entangled state Ψ is complete as soon as the particles have passed through the plate

The preparation entangles the particles so that at all subsequent times $p_a + p_b = -\Delta P$, and (only initially) $q_b - q_a = D$. Beller and Fine have pointed out (1994, pp. 14-15) that Bohr's experimental setup is in error because it cannot produce the special wavefunction (*) required by EPR. We cannot know $q_b - q_a$ when the particles later arrive at the apparatus: By this time the observed value q_a does not determine a particular value q_b.

- The measurement of ΔP is part of the preparation

It tells us the specific entanglement that has been achieved. After doing this, should we subsequently measure the momentum p_a of particle a, finding the specific value p_{a0}, then according to EPR (but not to Bohr), there is an element of reality p_{b0} corresponding to particle b. Everyone (EPR and Bohr) agrees on the correctness of quantum mechanics as a predictive tool; so that if we indeed subsequently measure the momenta of both particles, then we will find – as an experimental result – that invariably $p_{a0} + p_{b0} = -\Delta P$.

Conclusion of Bohr's response to EPR

Bohr explains why either type of measurement at either location will inevitably cause a minuscule kick to the base. He now says:

> Of course there is in a case like that just considered no question of a mechanical disturbance of the system under investigation during the last critical stage of the measurement procedure. But even at this stage there is essentially the question of *an influence on the very conditions which define the possible types of predictions regarding the future behaviour of the system.*
> (p. 700, original emphasis)

Commentary on Bohr's paper

Several authorities have remarked on the obscurity of the last statement. Arthur Fine (1986, pp. 34-5) accuses Bohr of lapsing "into positivist slogans and dogmas." He also believes that Einstein and his co-authors have forced Bohr to make an important concessionary change in his position: Previously Bohr had always insisted that changes in the physical situation were always caused by physical disturbances; but now Bohr was forced to change to "what one might call a doctrine of semantic disturbance" (Fine, p. 35). Measurement of one particle simply precludes meaningful talk of measurements of the other particle, despite the admitted lack of mechanical disturbance (Fine, p. 35). Tim Maudlin (2014, pp. 13-14) accuses Bohr of trying to have it both ways on the question of locality. He says that Bohr's italicised words are "simply incomprehensible."

My view of Bohr is slightly more sympathetic. I believe he is saying something meaningful although I do not agree with his position. According to the Copenhagen interpretation, quantum systems only have properties in the context of measurement. Moreover, these properties only come into existence once the measurement has been registered on a device, say when a pointer comes to rest, or a Geiger counter clicks. (This is one possible interpretation of his italicised words, which mention *"predictions regarding the future."*) At the time when Bohr's paper was written, measuring devices took of the order of about a thousandth of a second to operate. But light could cross a laboratory bench in roughly 30 trillionths of a second. Therefore, Bohr's position might be that, well before the measured values at A and B came to be registered on the relevant apparatus, the systems were no longer isolated. He devotes some space (pp. 700-1) to energy-time uncertainty and how this affects the impossibly fine timings required by his experimental setup; this is further evidence for my suggestion.

Thus, Bohr wrongly assumed that laboratory experiments could never achieve EPR non-locality. He could not have anticipated the future development of ultra-fast optical switches that would, decades later, enable a practicable experiment.

For Bohr, a **mechanical** disturbance must be something that is capable of being measured directly. The uncontrollable and unmeasurable disturbance of the experiment's base therefore would not have been considered by him to be 'mechanical'. Instead, he regarded it as something different, *"an **influence** on the very conditions."* Although Bohr does not explicitly say so in the paper, it is certain that he would also have denied that the concept of **indirect** measurement has any meaning.

Schrödinger's response to EPR

Schrödinger published no fewer than three responses to EPR, all of which are superbly instructive:

The first (1935a) paper

Schrödinger explained the curious non-local behaviour of quantum systems in terms of what he called "entanglement". His crucial discovery of this astonishing characteristic of our universe was directly inspired by his reading of the EPR paper. Schrödinger made the following definitions: Two systems with wavefunctions ψ^1 and ψ^2, with respective variables x and y, are *unentangled* if the combined system can be expressed as the product

$$\Psi(x, y) = \psi 1(x)\ \psi 2(y)$$

Otherwise, the systems are *entangled*. Here x and y are each shorthand for any number of coordinates in each of their respective systems (1935a, p. 556).

Schrödinger calls attention to what EPR have shown, namely that if a pair of particles are allowed to interact and then fly apart, they remain entangled – recall that in EPR the wavefunction Ψ before any measurement has the form of a *sum* of terms. The subsystems become unentangled only when a measurement is made on one of them and Ψ collapses to a single term, which indeed has the form of a product as in the equation just above.

Schrödinger comments on the non-locality that is a consequence of entanglement:

> It is rather *discomforting* that the theory should allow a system to be steered or piloted into one or the other type of state at the experimenter's mercy in spite of having no access to it. This paper does not aim at a solution to the *paradox*, it rather adds to it, if possible.
> (1935a, p. 556, my emphasis)

Note that it was non-locality by means of entanglement that Schrödinger regarded as discomforting and paradoxical. He also called the collapse of the wavefunction and its consequent disentanglement "sinister" (p. 555). In his analysis of the EPR thought experiment Schrödinger makes a striking analogy (a couple of trivial changes in notation have been made):

> Yet since I can predict *either* q *or* p without interfering with system B and since system B, like a scholar in an examination, cannot possibly know which of the two questions I am going to ask it first: it so seems that our scholar is prepared to give the right answer to the *first* question he is asked, *anyhow*. Therefore he must know both answers; which is an amazing knowledge, quite irrespective of the fact that after having given his first answer our scholar is invariably so disconcerted or tired out, that all the following answers are "wrong."
> ... Now I wish to point out that system B (say) has further knowledge. It does not only know these two answers but a vast number of others, and that with no mnemotechnical help whatsoever, at least none that we know of.
> (p. 559, original emphasis)

He goes on to prove this latter claim. The final part of the paper has relevance to Bohr's attempt to evade EPR's conclusion:

> But it is necessary to consider [my argument] lest one should believe that the antimonies could be solved by suggesting or proving that some of the observations must take a certain minimum time. Provided that they *relate* to a definite moment, this will not help us. It cannot be argued that, before the results are reached, the situation to which they refer has passed away. A prediction for time zero does not dissolve to nought as time goes on...
> The paradox would be shaken, though, if an observation did not *relate* to a definite moment. But this would make the present interpretation of quantum mechanics meaningless, because at present the *objects* of its predictions are considered to be the results of measurements for definite moments of time.
> (p. 562, original emphasis)

Schrödinger's conclusion would have greatly discomfited Bohr.

The follow-up (1936) paper

In this, Schrödinger strengthens the above result by proving that it is occasionally possible for the experimenter at location A to drive the system at distant location B into any state he chooses (!); whereas an experimenter local to B – who can measure the system directly – cannot do this. The proof makes instructive use of John von Neumann's concept of a 'mixture' of quantum states, and I do not have space to discuss it here.

Towards the end he affirms that quantum mechanics as a theory predicts non-locality "as a necessary and indispensable feature." He goes on to pose the really crucial question as to whether non-locality is true *in the Natural World*, and he says, "I am not satisfied about there being sufficient experimental evidence for that." He tentatively suggests a modification to quantum mechanics that would eliminate its non-locality property, without changing its predictions in the domain in which it had so far been tested; and he proposed that there might be other possible changes to the theory that could do the same (pp. 451-2).

Schrödinger's attitude to non-locality is summed up in the closing paragraph of this paper:

> It is suggested that these conclusions, unavoidable within the current theory but *repugnant* to some physicists *including the author*, are caused by applying non-relativistic quantum mechanics beyond its legitimate range. An alternative possibility is indicated.
> (p. 452, emphasis added)

The survey paper (1935b)

In November/December 1935 Schrödinger published in three parts a thoroughgoing survey of the then-current situation in quantum mechanics. He states in a footnote that it was inspired by EPR. Here I will use the canonical translation of this paper from German into English (Trimmer, 1980).

Schrödinger affirmed that the world in which we find ourselves is real in the sense that it exists independently of us. The project of science is to try to provide us with a mathematical model of this world that can be verified by observation and experiment. He gives the simple, classical model of the hydrogen atom, as consisting of two charged masses. There are several ways of expressing the variables of this system. For example, the components of the positions and momenta could be chosen. All of these could be specified independently, giving the state of the system at a given time (1980, pp. 323-5).

The theory of quantum mechanics differs in an essential respect from classical physics in that it allows only **half** of these variables to be chosen freely (say the components of position). Once these have been specified, no more can be said about the state of the system. On the basis of experimental evidence, Schrödinger remarks that this novel feature is unlikely to disappear in any future theory (p. 324).

Mathematical models are either attempts to represent the world that we find ourselves in (as Schrödinger and EPR contend), or they are attempts to represent our experiences and measurements (as Bohr and the Copenhagen School contend) (pp. 323, 328). In either case, all such models are the imagined creations of fallible humans. In particular, ψ is "an imagined entity" (p. 327). Schrödinger is having fun here: this central concept of quantum mechanics began as a creation of his own mind! Theories of physics should not be expected to be either complete or perfectly correct (pp. 324, 328). Here Schrödinger is clearly supporting Einstein over the contention of Bohr that quantum mechanics is necessarily complete.

Schrödinger next asks what the wavefunction can tell us about the character of the world. Could the wavefunction describe a world that consists of ensembles of systems of classical particles? Schrödinger proves that it cannot (pp. 326-7). He then asks: Could the wavefunction describe a world whose constituents are in truth occasionally smeared out or blurred? When we are discussing the atomic level, then there is no particular evidence on which we can decide this question one way or another. There now follows a famous thought experiment:

Schrödinger's Cat: Inside a box is placed a radioactive source, so tiny that there is a 50-50 chance of a particle being emitted within an hour. In addition, apparatus is arranged so that, if this event occurs, poison will be released. A cat is sealed inside this box for an hour. When the box is opened, the cat is found either alive or dead.

Schrödinger says, "The ψ-function of the entire system would express this by having in it the living and the dead cat (pardon the expression) mixed or smeared out in equal parts" (p. 328). Schrödinger did *not* regard this as a paradox. There is no reason one would expect ψ to give a complete picture of nature. The evidence of our eyes and our knowledge of cats easily trumps the absurdity of the human-made mathematical picture. One glance tells us not only that it is alive now, but also that it must have been so throughout its enforced sojourn in the box. (And a different account can be given about what happened in the box if the cat is discovered to be dead.) So ψ very obviously gives us an incomplete picture of what has been going on inside the box. He goes on to explain this situation in an analogy, likening quantum mechanics to a camera, and ψ to the image obtained: "There is a difference between a shaky or out-of-focus photograph [ψ, of a well-located cat] and a snapshot of clouds and fog" (p. 328). So, the reality of the cat is analogous to the circumstances of the first photograph. The second photograph is an absurd Copenhagen fantasy. To sum up, Schrödinger denies that the cat is somehow smeared out or in a superposition of two states. He also affirms that it is meaningful to discuss what must have happened to it while it was locked in the box, away from the possibility of observation. Any paradoxical assertions derive from a dogmatic adherence to the Copenhagen interpretation.

Throughout the paper Schrödinger expresses sharp criticisms of the Copenhagen interpretation. First there is its denial of the reality of the world: "We are told that no distinction is to be made between the state of a natural object, and what I know about it, or perhaps better, what I can know about it if I go to some trouble" (p. 328). Taken literally, this claim means that a system does not even have a state unless someone bothers to measure it (p. 332). Second, the Copenhagen interpretation's insistence that ψ gives a complete description means that the theory is dictating "which measurements are in principle possible" (p. 328). This is contrary to the spirit of science, in which we make any measurements we want and allow these – whatever they may turn out to be – to dictate the theory.

Schrödinger's survey includes a thorough discussion of quantum mechanics, entanglement, the EPR thought experiment, and what would later become known as "the measurement problem" (p. 329). It strengthens and generalises the arguments made in his two other responses to EPR (1935a and 1936) already discussed. The final sections of the survey speculate that the exact specification of time assumed by the theory is incorrect (under any interpretation such exactness is simplistic and clearly unattainable); and that this feature of quantum mechanics will need to be amended, in order to improve the theory and make it compatible with relativity. Over eighty years have passed, and so far no one has succeeded in doing this.

Bohm's response to EPR

Many years later, in his excellent book, *Quantum Theory*, David Bohm gave a detailed and careful response to the EPR paper in which he introduced several new ideas (Bohm, 1951, pp. 611-22). At the time of writing, he was a supporter of the Copenhagen interpretation in which the universe must be divided into the classical world in which the apparatus resides, obeying classical laws, and the quantum system under investigation, which obeys different laws. Properties of quantum systems only exist in the context of a particular experimental layout. He was, however, open-minded about quantum mechanics; he used his best endeavours to understand the quantum world, and he was alive to the possibility that science might develop beyond quantum mechanics:

> We may probably expect that even the more general types of concepts provided by the present quantum theory will also ultimately be found to provide only a partial reflection of the infinitely complex and subtle structure of the world. As science develops, we may therefore look forward to the appearance of new concepts, which are only faintly foreshadowed at present, but there is no strong reason to

suppose [that these novel concepts will be quasi-classical]. (Bohm, 1951, p. 622)

Bohm begins by restating (1) the EPR definitions of the completeness of a physical theory and (2) EPR's sufficient criterion for an "element of reality". He says that there are two further assumptions that EPR make implicitly:

(3) The world can be correctly analysed in terms of **distinct and separately existing** "elements of reality."
(4) Every one of these elements must be a counterpart of a *precisely* defined quantity appearing in a *complete* theory

(p. 612, my bold emphasis, original italics)

These are very subtle points. Recall that **elements of reality** are features of the universe which we try to obtain some grip upon by means of our **theories**. These additional assumptions together assert that there is a simple association between elements of reality and quantities in the theory, namely that of *pairing*. For example, these assumptions state that momentum p in the theory must be paired with some element of reality, call it P. Bohm claims that EPR are wrong to make these implicit additional assumptions, and hence wrong about pairing.

Bohm's solution to the EPR paradox is that, according to quantum mechanics "the wavefunction (in principle) can provide the most complete possible description of the system consistent with the actual structure of matter" (p. 620). But this description is not in terms of a pairing. Instead, "the properties of a given system exist, in general, only in an imprecisely defined form ... they are not really well-defined properties at all, but instead only *potentialities*, which are more definitely realised in interaction with an appropriate classical system, such as a measuring apparatus" (p. 620, emphasis added). Bohm gives the example that one may choose to measure either the position or momentum of an electron. He describes the wavefunction Ψ as a "mathematical abstraction" (p. 622).

Bohm gives a much more practicable example, based on measurements of spin, which fulfil the requirements of the EPR experiment (p. 614). This will be explained in the next section. The bare bones are that a pair of particles is emitted in opposite directions from a source, moving to distant locations A and B. Their entangled quantum state is such that a measurement of spin-direction made on the particle at location A instantly changes the quantum spin-state of the other. As Bohm says:

> Moreover, because the potentialities for development of a definite spin component are not realised irrevocably until interaction with the apparatus actually takes place, there is no inconsistency in the statement that while the atoms are still in flight, one [experimenter at A say] can rotate the apparatus into an arbitrary direction, and thus choose to develop definite and correlated values for any desired spin component of each atom [at both locations A and B].
> (p. 621-2)

Bohm also says:

> [The spin variant of the EPR-type experimental situation discussed here] actually permits us to measure a given observable without in any way disturbing the associated system.
> (p. 613)

These two statements taken together clearly show that Bohm accepts, and is not-at-all troubled by, non-locality. His denial of disturbance only makes sense because he denies any reality to the wavefunction.

Bohm's change of heart

Soon after this book was published, David Bohm changed his views radically and developed his own interpretation of quantum mechanics, called *Bohmian mechanics*, which has the following properties:

- It is identical to quantum mechanics in terms of the predictions it makes

In particular, it is certain that no experiment could ever differentiate between the Copenhagen interpretation and Bohmian mechanics (at least in situations where one may ignore relativity).

- It contains elements of reality corresponding to the positions of particles

So particles are physically real and have definite trajectories.

- It is non-local and deterministic
- The wavefunction is physically real and guides the particles

Bohmian mechanics will be discussed in more detail in the next chapter.

The remainder of this chapter will show how ideas such as entanglement and non-locality – originally presented as theoretical abstractions – later came to be tested experimentally.

Bell's inequality

For many decades, the arguments between EPR and Copenhagen were inconclusive: you might for instance affirm the validity of indirect measurements, and your opponent could do no more than contradict you. Then, in 1964, John Bell (who was an uncompromising realist, and thus was opposed to the Copenhagen interpretation) published a brief paper entitled *On the Einstein Podolsky Rosen Paradox* that would eventually settle this dispute. He concluded that the crucial feature that is in question with regard to quantum mechanics, is whether it is a local or non-local theory. Moreover, he devised a principle that would allow experimental tests of this disputed feature to be carried out (Bell, 1964, p. 199).

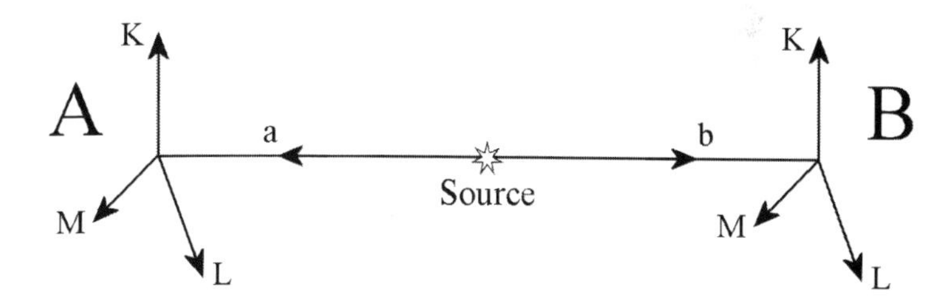

Figure 5.3: Layout of a proposed experiment to test Bell's inequality. Pairs of spin-½ particles a, b are emitted from the source and travel to distant locations A, B. Apparatus at each location can be oriented to measure the spin of its particle in one of the directions K, L, M, which are at 120° to one another, and perpendicular to the line of travel of the particles.

In his 1964 paper Bell proved a key theorem now called *Bell's inequality*. This states that if locality holds in our universe, then a certain inequality must be satisfied. Recall that locality means (roughly) that events or states of affairs at location A can have no influence that would change events or states of affairs at distant location B. I will not give Bell's inequality in general, but just one fairly straightforward example of it, which is due to David Mermin (see Lewis, 2016, pp. 34-37). Figure 5.3 shows a possible experimental setup. Pairs of spin-½ particles (for example electrons) are emitted from a source and travel to targets at widely separated locations A and B. The spins of the particles are measured in one of the three directions K, L, M, all perpendicular to the line of travel of the particles, as shown in the Figure. The experimenters at each location separately make a free choice of this direction; alternatively, the direction of measurement can be chosen at random. The experimenters are isolated from each other and working independently.

Consider the situation at one of the targets. The measuring device – a Stern-Gerlach apparatus – has been set up in one of the directions, and the spin of a particle is measured. Recall that for a spin-½ particle there are only two possible outcomes: either the particle is found to spin in line with the orientation of the apparatus – we will say that the particle's spin is 'up'; or it will be found to spin in the opposite direction, and this result we will call 'down'.

When the entire experiment is carried out for multiple pairs of particles, quantum mechanics makes two predictions about the results. The **first** is that when both measurements happen to be made in the same direction (say K), then the spins must disagree: if particle a is spin up, then particle b must be spin down or vice versa (by conservation of spin). **Second**, when the measurements happen to be made in different directions (say the apparatus at A is oriented towards K, and that at B is oriented towards L), then the results will agree – either both up, or both down – with probability 0.75 (and so of course they will disagree with probability 0.25).

Sketch proof of the second prediction: Let $|r\uparrow\rangle$ correspond to direction K; and let $|z\uparrow\rangle$ correspond to direction L. The angle θ between K and L is 120°, so $\theta/2 = 60°$. From the previous chapter we know how to express $|r\downarrow\rangle$ in terms of $|z\uparrow\rangle$ and $|z\downarrow\rangle$:

$$|r\downarrow\rangle = \sin(60°)\ |z\uparrow\rangle - \cos(60°)\ |z\downarrow\rangle$$

Before we measure the spin of the particle at A, we have no idea as to what its spin state is. After the measurement in the K direction, we know that its spin will be either $|r\uparrow\rangle$ or $|r\downarrow\rangle$. Suppose it is $|r\uparrow\rangle$, then we know that the spin state of the particle at B is the opposite; i.e., $|r\downarrow\rangle$. The probability of finding the particle at B spinning up in the L direction (i.e., spinning $|z\uparrow\rangle$) is $\sin^2(60^\circ) = 0.75$. This, then, is the probability that both particles will be found to be spinning up.

Suppose instead, the particle at A is found to be $|r\downarrow\rangle$ when measured. In this case, the particle at B will be in the state $|r\uparrow\rangle$. Recall that this can be expressed as:

$$|r\uparrow\rangle = \cos(60^\circ)\ |z\uparrow\rangle + \sin(60^\circ)\ |z\downarrow\rangle$$

The probability of agreement (this time both down) is again $\sin^2(60^\circ) = 0.75$ **Proof ends**.

Bell uses proof by contradiction. He makes a provisional assumption that **our universe is EPR local**, and demonstrates that this assumption contradicts the two predictions above. We begin by making this assumption now.

Suppose we make a preliminary test, orienting both instruments in direction K. By observing the result at A, we can predict with certainty the result at B. From our **assumption of locality**, after a little thought, we can conclude that the result at B must be predetermined by some hidden factor (a.k.a. 'element of reality') within the particle b. If the result at B is up, we will call the hidden factor 'u'; whereas if the result is down, we will call the hidden factor 'd'. By the first prediction above, if particle a has factor u, then we know particle b has factor d (and vice versa).

We make a second preliminary test, this time making measurements in arbitrary directions. In this new situation, according to the second prediction, the agreements are only statistical. This means that the particles must have separate hidden factors, specific to each orientation of their device. (If there were only a single factor, common to all orientations, then the results would **always** disagree.)

Figure 5.4 shows all possible combinations of hidden factors for a given pair of particles. There are eight possible cases, which are of two types:

1) *All hidden factors the same*

CASE 1, on the left of Figure 5.5, shows what happens if the hidden factors are all uuu for particle a, and hence ddd for particle b. We ignore runs of the experiment in which the devices happen to be oriented in the same direction. We see that there is invariably disagreement as to the results obtained: particle a is always up, and particle b is always down. Case 8 is similar to this: Again, there is complete disagreement, but now a is always up, and b is always down. To sum up: in cases 1 and 8 the probability of agreement is 0.

2) *Not all hidden factors the same*

CASE 2, on the right of Figure 5.5, shows what happens if the hidden factors are not all the same. Again, we ignore runs of the experiment in which the devices happen to be oriented in the same direction. From this Figure it can be seen that the probability of agreement is 0.66. Cases 3 through 7 are similar to case 2. To sum up: in cases 2 through 7 the probability of agreement is 0.66.

Particle:	**a**			**b**		
Direction measured:	**K**	**L**	**M**	**K**	**L**	**M**
CASE 1	u	u	u	d	d	d
CASE 2	u	u	d	d	d	u
CASE 3	u	d	u	d	u	d
CASE 4	u	d	d	d	u	u
CASE 5	d	u	u	u	d	d
CASE 6	d	u	d	u	d	u
CASE 7	d	d	u	u	u	d
CASE 8	d	d	d	u	u	u

Figure 5.4: Possible values (ud) of the proposed hidden factors in the Aspect-type experiment sketched in the previous Figure. There are eight possible cases labelled 1 through 8.

CASE 1		**Particle a** K u	L u	M u
Part-	K d	■	×	×
icle b	L d	×	■	×
	M d	×	×	■

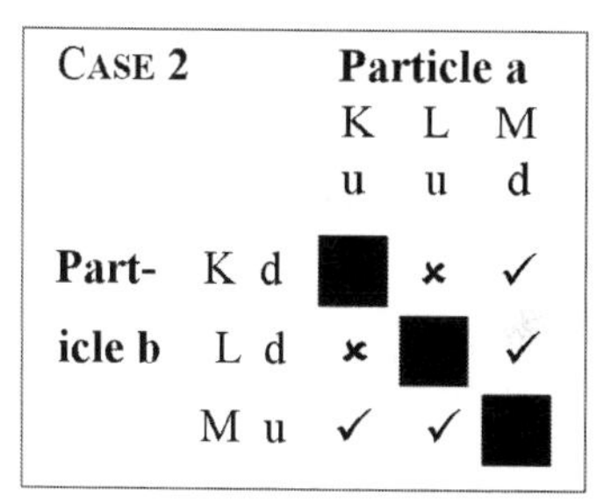

*Figure 5.5: **CASE 1** on the left shows that there is zero agreement when the hidden factors are the same for all directions KLM. (CASE 8 is similar.) **CASE 2** on the right shows that there is ⅔ agreement when the hidden factors are NOT the same for all directions KLM. (CASES 3 through 7 are similar.)*

From the discussion of these Figures, it can be seen that the agreement between observations of the particles is **at most** ⅔ when the orientations of the measuring devices differ. This is an example of a Bell inequality. But, as noted earlier, the probability that quantum mechanics gives is 0.75. If the mathematical formalism of quantum mechanics is correct then Bell's inequality must be broken. This would not have come as a surprise to Bell. The inequality was explicitly derived on the assumption that locality is true of our universe. As Bell well understood, his inequality provided a practical test that would enable experimenters to discover whether or not this assumption about our universe is true (Bell, 1964, p. 199).

Experiments by Alain Aspect and others

Bell's 1964 "inequality" paper lay fallow for many years, but by the late 1970s experimentalists had begun the attempt to put it to the test. Preliminary findings were ambiguous but later results were in agreement with the predictions of quantum mechanics; showed that Bell's inequality was broken, and hence were in favour of non-locality as a feature of the universe. (This should not be interpreted as necessarily supporting the Copenhagen interpretation. The experiments supported the quantum recipe – i.e., bare-bones quantum mechanics, absent any mention of the theory's interpretation.)

These early experiments had the fault that the measurement apparatus at locations A and B had to be moved into position by hand before each observing run. There was thus ample time for light to travel between A and B in order to "pass on, to particle b, information about the experimental setup at A" (and vice versa). Whilst a mechanism for such information-passing seemed implausible in the extreme, it remained a loophole in the conclusion of non-locality. Alain Aspect devised an experiment that would circumvent this difficulty and, when he approached Bell about his idea, he was received with great encouragement and enthusiasm.

Figure 5.6: Layout of the Aspect (ADR) experiment to test non-locality. Photons a and b are emitted from the source and travel to the two locations A, B. At location A, switch S_A will be used to deflect photon a to one of the fixed polarizers K or L. These latter measure polarization in different orientations. Switch S_B works similarly. The switches make their choices so rapidly that regions A and B are spacelike separated.

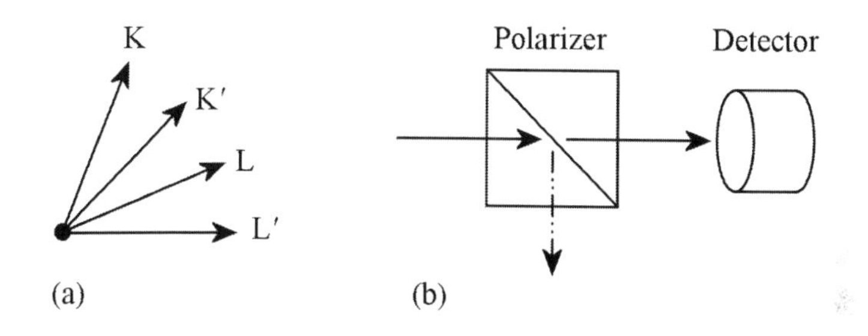

Figure 5.7: Details of the ADR experiment. (a) The fixed relative orientations of the four polarizers. Each angle (e.g., KK′) is 22½°. The dot (•) marks the line of travel of a photon perpendicular to the page. (b) Each polarizer consists of two prisms. In this experiment, only photons passing straight through the polarizer are recorded. Reflected photons are ignored.

The first experimental test of Bell's inequality with regions A and B spacelike separated was carried out by Aspect and his team, ADR (Aspect, Dalibard & Roger, 1982). See Figures 5.6 and 5.7. (There were some relatively minor changes from the example of Figure 5.3.):

- Photons were used as test particles; and polarization was measured rather than spin
- At each location, A and B, only two different orientations of the measuring apparatus (the polarizers) were needed, rather than three

In order to allow for this latter simplification, the ADR experiment made use of a more practicable variant of Bell's inequality, called CHSH after its devisers Clauser, Horne, Shimony and Holt (1969). The CHSH inequality depends upon the same fundamental principles as Bell's, and the mathematics of it will be omitted here.

- The orientations of the polarizers K′, L′ at B are different from the corresponding orientations of the polarizers K, L at A

This is another consequence of using the CHSH inequality. The orientations are shown in Figure 5.7(a). We imagine that the initial paths of photons a, b are into and out of the page respectively. This is marked by the dot (•). A polarizer shown is slightly more detailed in Figure 5.7(b). It consists of a pair of face-to-face prisms. Photons can either pass directly through the polarizer to be recorded by a photomultiplier tube. Or they can be reflected through a right angle by the touching faces, as shown by the dotted line. (Reflected photons are not counted in this experiment.)

The crucial differences in ADR's experiment are as follows. Instead of physically moving a polarizer between K and L (or K′ and L′):

- Polarizers, with an orientation that is **fixed** throughout the experiment, are set up at **every** location (K, L, K′ and L′)

In Figure 5.7(b), the orientation of each polarizer is fixed in position – once-and-for-all before the experiment.

- A super-fast optical switch directs photon a either to K or L. A similar switch directs photon b either to K′ or L′.

The speed of light is very close to 0.3 metres per nanosecond (per billionth of a second). The switches are 12 metres apart and they are tuned to switch approximately once every 20 nanoseconds. During this time, light can only travel 0.3×20 = 6 metres, so the switches, and hence the regions A and B, are spacelike separated.

To sum up: The ADR experiment is sketched in Figures 5.6 and 5.7. Photons are emitted from the source and travel towards regions A and B. Switch S_A acts on photon a, causing it to be measured: either by polarizer K, which is fixed in a certain orientation (also called K); or by polarizer L, which is fixed in orientation L. Switch S_B operates similarly on photon b: directing it either to polarizer K′ in orientation K′; or to polarizer L′ in orientation L′.

A great deal of detail has been omitted in this sketch: the efficiency of the detectors, the electronics needed to record the data, and much else. The conclusion of the experiment was in line with the predictions of the quantum recipe, broke the CHSH inequality, and contradicted the idea that the universe is local.

As ADR themselves pointed out, there were some loopholes in the technical design. The switches were periodic (at slightly different rates, both very close to 20 billion times per second). This left open the in-principle possibility that the photons could somehow use these periodicities to 'predict' which measurement would be made (not that any physicist has ever suggested a remotely plausible mechanism by which this might happen). In 2018 Dominik Rauch, leading a team of many others, closed this gap, by using fluctuations in the light from two distant quasars to perform in effect the role of the switches (Rauch *et al.*, 2018). A considerable amount of experimental effort still continues in order to close ever more fantastical loopholes. Experiments have now been carried out with regions A and B separated by several tens of kilometres.

Conclusions we may draw about Aspect-type experiments

What conclusions can we draw from the results of these experiments? The answer is very subtle, but Travis Norsen has given an extremely succinct and lucid explanation, which I can do no better than to repeat here (I've slightly abbreviated the final sentence):

> Bell's theorem, then – taken here to mean the proof that local hidden variable theories are wrong – must be understood as the second part of an overall two-part argument, the first part of which is the EPR argument. Schematically, the two-part argument goes like this:
>
> **EPR:** locality → X
> **Bell:** X → conflict with experiment
>
> Here "X" stands for something like "local deterministic hidden variables", but somehow the logic is easier to grasp by suppressing this. Obviously, if X implies conflict with experiment then X must be false, which means we cannot maintain locality (because locality entails X!).
> (Norsen, 2017, pp. 233-4)

There are thus two conclusions:

- The universe is non-local
- Taken in its entirety, X is false

There are several ideas packed into assertion X, and some variant Y of X might be true. For example, Y = '**non**-local deterministic hidden variables' remains a possibility.

Misconceived conclusions

Misconceptions about what EPR, Bell and Aspect proved are still shockingly widespread. Here are some of the most common:

- *The Aspect experiment proves that no deterministic account can be given of microphysical behaviour*

This is false, as Bohmian mechanics proves.

- *The Aspect experiment proves that quantum mechanics necessarily gives a complete picture of the universe*

This is false, as Bohmian mechanics, which contains hidden variables, proves.

- *The Aspect experiment vindicates the Copenhagen interpretation*

This is false because Bohmian mechanics makes identical predictions to Copenhagen.

The misrepresentation of Einstein

These experiments have also been used to cast aspersions on Einstein. This still occurs today – even at the most prestigious universities. Now that the experiments have been discussed, it is worthwhile reiterating some of the myths:

- *Einstein rejected quantum mechanics and its experimental results*

Einstein accepted all the results of quantum mechanics – it was the irrealism and vagueness of the Copenhagen interpretation that he objected to.

- *Einstein was mistaken in assuming that the universe is local*

This is true but highly misleading. Einstein was the first to see that quantum mechanics implies that: either the Copenhagen interpretation is incomplete; or that the world is non-local. He and his co-authors explicitly put locality into the EPR paper as a hypothesis. It is perfectly honourable for a scientist to make a plausible, explicit, testable proposal that is later shown to be wrong. Moreover, the idea proved fruitful because it led directly to the experimental verification of entanglement and non-locality.

In contrast, Bohr's response to the EPR paper was to show that non-locality could not be achieved in a specific experimental setup on a laboratory bench. He (wrongly) went on to argue at length that **no** experimental setup could possibly detect non-locality. Bohr's philosophical attitude would have been that what is impossible to measure cannot exist. So, during that era, Bohr held that locality was a necessary truth and non-locality was meaningless. Moreover, Bohr refused to develop intuitions about is going on in the quantum world: any attempt at understanding was anathema to him, and he strongly discouraged it in his followers. Because he rejected the creative role of intuition, Bohr's ideas were necessarily obscure and thus unfruitful.

- *Einstein was wedded to determinism*

Einstein preferred deterministic theories, but this was not a paramount issue; he was willing to consider stochastic theories if the facts required this. If one believes, as everyone did during that era, that the universe is local, then from this assumption – as the EPR paper demonstrates – **determinism follows as a logical consequence**. Determinism is not an assumption of the EPR paper.

- *Non-locality and the results of the Aspect experiment were foreseen by Bohr and other followers of the Copenhagen interpretation*

Non-locality (in terms of entanglement) was first discussed by Schrödinger, a critic of Copenhagen (1935a, b; 1936). Its experimental discovery was the joint work of Copenhagen dissidents EPR, Bell, Clauser and Aspect. This, the most bizarre feature of the quantum world, was initially greeted with astonishment, but almost immediately accepted (one might say purloined) with complacency by Bohr and company. Bohr certainly did not anticipate non-locality.

Chapter 6

Interpretations of quantum mechanics

Chapter 4 introduced quantum mechanics as a recipe – a scientific methodology that, given an experimental situation, makes astonishingly accurate predictions concerning the outcome that will be observed. As a predictive tool it is unsurpassed. But what – if anything – does the theory tell us about what is going on in the world? This chapter sets out to answer this question. An *interpretation* of quantum mechanics is a set of ontological claims that are appended to the bare predictive tool, which change it into a *science* – a system that gives us some insight, invariably limited to a greater or lesser extent, concerning the nature of the universe.

The chapter describes four representative interpretations from the more than a dozen that are extant; namely Copenhagen, pilot-wave, many-worlds, and GRW. It ends with a discussion of what characteristics an interpretation should have – in my view – which will enable it to form part of an adequate approach to the mind-body problem.

The Copenhagen interpretation

Recall from the previous chapter that this position is, broadly speaking, positivist in tenor. It claims that our understanding of the quantum world is at best extremely limited. The name "Copenhagen interpretation" is itself controversial. According to Becker (2018, pp. 61-62) the expression was coined by Heisenberg as recently as 1955. Before that, terms such as "the Copenhagen spirit" were occasionally used. It was by no means one single position, but each leading light within or close to Copenhagen-thinking had their own distinctive ideas. Here I will sketch the positions of Bohr, Heisenberg, Peierls, and von Neumann.

Niels Bohr

Bohr's philosophical views have already been discussed extensively. Among them were:

1. The sole criterion for the truth of quantum mechanics is agreement with experiment
2. "Irrealism"

Bohr was ambiguous about this: Usually he asserted vigorously that physics could tell us nothing – even in principle – about the quantum world, outside of the context of performing a measurement. But occasionally he claimed that uncontrolled, irreversible interactions between a quantum system and its environment occur over the course of time – irrespective of the whether or not the system is being observed. (He called such an interaction a "quantum of action".)

3. Complementarity

Bohr insisted that it is impossible – even in principle – to form a single consistent picture of what is actually going on when we investigate the quantum system. The most lucid conceptual picture we could possibly imagine will never be better than a patchwork of contradictory ideas.

Werner Heisenberg

Heisenberg was also a key member of the Copenhagen school. His development of matrix mechanics (in contrast to Schrödinger's wave theory) was specifically aimed at restricting science to what humans could actually observe when performing experiments – and to eliminate any intuition about (or mention of) what is actually going on in the world. Paul Dirac was later to show that Heisenberg and Schrödinger's mathematical formalisms, though at first sight wholly dissimilar, were in fact equivalent.

Much later in life, Heisenberg, while nominally remaining a follower of Copenhagen, reached a more objective understanding of quantum mechanics by introducing the concept of an interaction of a quantum system with a measuring device, which exists in the world prior to any act of human observation. Such an objective interaction, I will call an *actual event* for short (following Stapp's terminology, 2004a, p. 126). Heisenberg's concept of an actual event may be seen in the following:

> The observation itself changes the probability function discontinuously; it selects of all possible events the actual one that has taken place … the transition from the "possible" to the "actual" takes place during the act of **observation**. If we want to describe what happens in an atomic event, we have to realise that the word "happens" can only apply to an observation, not to the state of affairs between two observations. **It applies to the physical not to the psychical act of observation**, and we may say that the transition from "possible" to "actual" takes place **as soon as the interaction of the object with the measuring device, and thereby the rest of the world, has come into play; it is not connected with the act of registration of the result in the mind of the observer**.
> (Emphasis added, Heisenberg, 1958, p. 22)

Thus, Heisenberg held that objective interactions a.k.a. actual events can take place in the absence of any mind. For example, in the early universe, a high-velocity particle might make a permanent scar in a quartz crystal – and it might never happen that someone comes along later to take a look at it. Heisenberg never made clear where and when such actual events would occur in space and time, but his notion foreshadowed the more precise ideas of GRW, to be discussed later.

Rudolf Peierls

Peierls had a distinguished career in physics, and remained throughout his life a strong defender of the Copenhagen interpretation. The quotations given here are all taken from an interview he gave (for Davies & Brown, 1993, chapter 5). Peierls objected to the term "Copenhagen interpretation". When asked why, he replied:

> Because this sounds as if there were several interpretations of quantum mechanics. There is only one. There is only one way in which you can understand quantum mechanics. There are a number of people who are unhappy about this, and are trying to find something else. But nobody has found anything else which is consistent yet, so when you refer to the Copenhagen interpretation of the mechanics what you really mean is quantum mechanics. And therefore the majority of physicists don't use the term; it's mostly used by philosophers.
> (See Davies & Brown, 1993, p. 71)

So Peierls states explicitly that the Copenhagen interpretation and quantum mechanics are one and the same. Because of this he finds the Aspect experiments uninteresting: for him, the results merely confirm validity of quantum mechanics, as expected (p. 77). It is curious that Peierls believes that no alternative consistent interpretation has been found. Later he says:

> There is no sensible view of hidden variables which doesn't conflict with these [Aspect's] experimental results. That was proved by John Bell, who has great merit in having established this. [...] So, I think the answer is that these experiments, at least, dispose of all hidden variables theories, but perhaps somebody can still come up with one which is compatible with these experiments.
> (p. 77)

The reference to "hidden variables" theory is to Bohmian mechanics, which will be discussed later. For now, it is sufficient to say that the predictions of Bohmian mechanics are identical to those of the Copenhagen interpretation. Bohmian mechanics is therefore **necessarily** in accord with the findings of Aspect's experiment. It is surprising that Peierls does not acknowledge this.

Peierls was an irrealist:

> I don't know what reality is [...] Again I object to your saying reality. I don't know what that is. [...] But if we try to maintain this Einstein ideal that there must be such a thing as reality, then we get into lots of logical troubles with quantum mechanics.
> (pp. 74, 75, 76)

Resulting from this, Peierls' attitude towards what we can say about the very early universe is interesting, and somewhat paradoxical:

> We can see around us in the universe many traces of what happened there before. [...] We can therefore set up a description of the universe in terms of the information which is available to us. ... [Because physics] consists of a description of what we see or what we might see and what we will see, and if there is nobody available to observe this system, then there can be no description.
> (p. 75)

One wonders what a typical evolutionary biologist or cosmologist would make of this. Fossil bones are traces that allow biologists to describe dinosaurs. But, on the basis of such evidence, can we assert without equivocation that dinosaurs once existed 'in flesh and blood'? Or are dinosaurs nothing more than descriptive theoretical constructs enabling us to explain present-day fossil traces? Biologists would accept the former and reject the latter alternative. Similar remarks hold for the status of the early universe as this is understood by cosmologists.

For Peierls, in contrast to Heisenberg, it is **essential** that every observation involves a human mind:

> Suppose you have an apparatus which tells you whether or not a radioactive atom has decayed by the position of a pointer on a dial. You could describe this apparatus in terms of conventional physics. But before you look at the apparatus there are two possibilities for what the result might be, and quantum mechanics will give you the probability that the pointer will be in one position or the other. ... The moment at which you throw away one possibility and keep the other is when you finally become *conscious* of the fact that the experiment has given one result.
> (Emphasis original, p. 73)

According to Peierls, even a camera or a computer, as a measuring or recording device, could never be an observer (p. 74). This brings us to a particular consequence of the Bell-Aspect experiment, which proved, as was shown in the previous chapter, that the universe **must** be non-local in its essential character, independent of any scientific theory that may be devised to describe it. Why did Peierls fail to be astonished by this? It remains open for Peierls to claim that there can be no evidence for non-locality – we cannot even discuss such features of the universe. How is this so? The results – the spins of the pairs of particles – were recorded on a computer during the course of the experiment. But nothing was **observed** until these were brought together, and then witnessed by a scientist. Measurements first came into being at a single location: that of the scientist.

Peierls' position shows how a consistently-held belief in the Copenhagen interpretation as a philosophical system leads one into incuriosity about the character of the physical world, in which a multitude of questions cannot even be asked.

John von Neumann

John von Neumann is famous both for his brilliant mind, and for a seminal work, *Mathematical Foundations of Quantum Mechanics*, published in German in 1932. The book is highly advanced mathematically, being the first account of quantum mechanics in terms of Hilbert spaces, which were lightly sketched in chapter 4. Experts agree that von Neumann's account was, by comparison, less sophisticated in terms of its physics. This is for the following three reasons.

First, as discussed previously, measurements are described in terms of operators. Although there are formal rules for associating each type of measurement (say momentum or position) with a different type of operator, these associations are ad hoc, and give little physical insight. Mathematical restrictions are placed on the type of operators that are permissible (these are called "Hermitian operators"); but the restrictions are merely sufficient to ensure that the predicted results of measurements are real numbers.

Second, von Neumann wrongly claimed to have demonstrated that no hidden-variables interpretation of quantum mechanics could be given (1932, pp. 323-25). His method of proof was to assume that some hidden-variables interpretation was given, and from this to reach a contradiction.

Although this proof was accepted by almost all physicists for many decades, John Bell eventually published a paper showing that the result was not watertight (1966, pp. 448-49). Von Neumann had made the naive physical assumption that, in the given experimental situation, the hidden variables must have the same mathematical characteristics as the manifest variables.

But hidden variables might be otherwise than he assumed. A readable account of von Neumann's mistake is given in Mermin, 1993, pp. 805-6. Mermin also points out that Grete Hermann, a mathematician and philosopher, had found this "glaring deficiency" in von Neumann's (1932) proof as early as 1935, but "she seems to have been entirely ignored" (p. 805).

Third, von Neumann claimed (1932, pp. 421-37) to have solved the measurement problem for all practical purposes, but this is somewhat exaggerated, as we shall see in the next subsection. As a preliminary to this, we must ask: To what extent does von Neumann regard the wavefunction and its collapse as being physically real? Usually, he discusses them in unambiguously physical terms, and without qualification. For example:

> The difference between these two processes [process 2 is evolution according to the Schrödinger equation; process 1 is the collapse of the wavefunction] is a very fundamental one: aside from the different behaviours in regard to the principle of causality, they are different in that the former is (**thermodynamically**) reversible, while the latter is not.
> (Emphasis added, von Neumann, 1932, pp. 351, 418)

In one place, in which he is quite discursive, he gives a careful statement of his attitude. (More typical of his writing is to give a few words of prose interspersed within a blizzard of formulae.):

> Let us assume that we do not know the state of a system **S**, but that we have made certain measurements on **S** and know their results. In reality, it always happens this way, because we can learn something about the state of **S** only from the results of measurements. More precisely, the states are **only** a theoretical construction, only the results of measurements are actually available, and **the problem of physics is to**

> **furnish relations between the results of past and future measurements**. To be sure, this is always accomplished through the introduction of the auxiliary concept "state," but the physical theory must then tell us on the one hand how to make from past measurements inferences about the present state, and on the other hand, how to go from the present state to the results of future measurements.
> (Emphasis added, von Neumann, 1932, p. 337)

Here von Neumann's statement about 'the problem of physics' is positivist in tenor.

The measurement problem

The measurement problem is a key difficulty of the Copenhagen interpretation, and attempting to solve it is a major driver of the search for alternative interpretations. It has several facets.

- Quantum mechanics gives no account **in physical terms** as to (a) what a measurement is or (b) under what circumstances measurements occur.

It is natural to hope that a physical theory would be able to give physical accounts of both of these things. The assertion (made by the Copenhagen school) that no such account is possible is a metaphysical claim, rather than a physical one. One might conceivably give philosophical arguments in favour of this assertion; but no empirical facts could substantiate it.

- Quantum mechanics does not specify in physical terms the boundary between a quantum system and its environment. This leads to ambiguities.

A quantum system is described in terms of its quantum state, ψ. Its environment (including measuring devices and the observer) can only be described – for lack of anything better – in classical terms. Von Neumann (1932, pp. 421-37) gives the following example. Suppose we want to perform a measurement (say taking the temperature) of some quantum system, using some device (say a thermometer). We can divide the world into three parts: part I, the system being measured; part II, the measuring device, in this case the thermometer; and part III, the rest of the universe, including the laboratory, and the observer. Typically, one would put the dividing line between the quantum world and the classical world between I and II; i.e., one would treat the thermometer as part of the classical world. Von Neumann was able to prove that exactly the same empirical results would be observed (the scientist would have the experience of seeing the same temperature readings on the thermometer) if the notional dividing line were instead placed between parts II and III. This latter dividing line has the effect of treating the thermometer as a quantum system, able to get into a superposition of states (just as was alleged to happen to Schrödinger's cat). The eye of the observer is now treated as the measuring instrument. This supposed dividing line is called the *Heisenberg-von Neumann cut*; sometimes abbreviated to the *Heisenberg cut*.

According to von Neumann, it is possible to move the cut even further, so that we can regard the observer's eye and optic nerve not as part of that person, but instead as part of the quantum system (1932, p. 420).

> But this does not change the fact that in each method of description the boundary must be put somewhere, if the method is not to proceed vacuously, i.e., if a comparison with experiment is to be possible. **Indeed, experience only makes statements of this type: an observer has made a certain (subjective) observation; and never any like this: a physical quantity has a certain value.**
> (Emphasis added, von Neumann, 1933, p. 420)

This shows that von Neumann regarded consciousness as essential to the concept of measurement. But what is actually going on physically with the thermometer (and the observer's eyes)? Different cuts give vastly different answers. The Copenhagen response is that one should not ask this question. Von Neumann's attitude seems to be that, because different cuts make the same predictions as to what outcome the observer will witness, there is nothing to be concerned about. John Bell was highly critical of this, calling the Heisenberg-von Neumann cut a "shifty split" (Bell, 1990, p. 34). Along with the two dissimilar physical descriptions of the thermometer, there are two distinct types of probability that are associated with it. If the thermometer is regarded as a part of the classical environment, then these probabilities are classical. But if the thermometer is regarded as part of the quantum system, then the probabilities are derived from the (amplitude of the) wavefunction, using the Born Rule. Von Neumann used these two types of probability to develop his concept of a *density matrix*, a generalisation of the wavefunction, which I do not have space to describe here.

- Does measurement necessarily involve conscious observers?

If Yes, then we are led to the paradoxes of Schrödinger's cat (or von Neumann's thermometer (and even the observer's eyes!)) having several bizarre – and mutually inconsistent – descriptions. If No, then we are led to question-begging and vague assertions as to when measurements occur. For example, a measurement occurs when a quantum system interacts with a 'macroscopic' system. Another would be to try to **physically define** a measurement as happening when a thermodynamically irreversible process occurs. (An earlier quote by von Neumann merely characterises measurements in terms of them having this feature. It does not attempt to physically account for when, where, or how measurement events occur.)

- The term "measurement" has misleading associations.

The word wrongly suggests that some pre-existing fact is being revealed by the measuring device. For this reason, Bell goes so far as to suggest that the word should be banned from quantum mechanics texts (1990, p. 34). As we shall see, even in hidden variable theories, such as de Broglie-Bohm below, a Stern-Gerlach apparatus does not reveal the pre-existing spin of a particle.

To sum up, Copenhagen, once held to be unquestionable, remains the dominant interpretation of quantum mechanics. It exists in many contrasting, contradictory – even self-contradictory – variants. Moreover, it is unacceptably hazy on many key issues – even putting limits on the questions that physicists are permitted to ask.

The pilot-wave interpretation

Pilot-wave theory is an interpretation of quantum mechanics in which there exist quasi-classical particles, which are steered by the wavefunction. It goes by several alternative names: *Bohmian mechanics* and the *de Broglie-Bohm interpretation*. It is the earliest alternative to the Copenhagen interpretation.

Louis de Broglie introduced pilot-wave theory as applied to a single particle for his doctoral thesis. Soon afterwards he presented his ideas at the 1927 Solvay Conference, and was roundly criticised, especially by Pauli. This led him to lose interest in his theory, and he never returned to it. In 1952 David Bohm developed pilot-wave theory for any number of particles. He also proved that for non-relativistic situations it gives exactly the same results as does the Copenhagen interpretation. Response to Bohm's theory was even more hostile, and often the criticism was not scientifically or rationally based; Robert

Oppenheimer is reported to have said, "If we cannot disprove Bohm, then we must agree to ignore him" (Peat, 1997, p. 133).

Key features

There are several features that determine the character of pilot-wave theory. The first is that it has a clear, precise **ontology** in which there exist two types of entity:

- Quasi-classical particles with well-defined trajectories

Particles follow definite paths through space. Be warned, however, that some properties of particles (notably their spins) still have no classical counterparts.

- The wavefunction Ψ of the entire universe

This is in contrast to Copenhagen, which can only meaningfully be defined on the particular quantum system that is under investigation. In pilot-wave theory, Ψ always obeys the Schrödinger equation without exception – it never collapses. A specific function, called the *quantum potential*, derived from Ψ, steers all of the particles in the universe, given all of their present positions – hence Ψ is regarded as a pilot-wave, by analogy to a pilot steering a ship.

We can use John Bell's term *beable* ("that which can exist" in the universe; as opposed to *observables* – "that which can be observed") to express the ontology. The positions of the particles are the *local beables* of the theory: they have a definite location in space. As Bell remarks, without at least some local beables, a physical theory has no empirical meaning. The universe's wavefunction is a *non-local beable*. In pilot-wave theory, everything within the ontology – both the particle trajectories and the universe's wavefunction – are expressed in terms of **position**.

The second feature is that the theory is **deterministic**. At any given time t_0: Ψ is an exact function defined over configuration space; and moreover, the positions of all particles have exact values. From this, Ψ and all particle positions are fully determined for all times t, both before and after t_0.

Finally, the theory is explicitly **non-local**. Suppose particles 1 and 2 come from a common source, and remain entangled, even though they are now far apart. Some apparatus measures 2. In the course of this physical process, the apparatus must introduce what Bell calls "analysing fields" near 2, which affect the quantum potential. Bell affirms that according to pilot-wave theory:

> The trajectory of 1 then depends in a complicated way on the trajectory and wavefunction of 2, and so on the analysing fields acting on 2 – however remote they may be from particle 1. So in this theory an explicit causal mechanism exists whereby the disposition of one piece of apparatus affects the results at a distant place.
> (Bell, 1966, p. 452)

In short, pilot-wave theory explains how the trajectories of distant particles are entangled. A Stern-Gerlach apparatus performing a measurement on particle 2 instantly affects the results of a similar measurement made on particle 1 (see also: Bohm & Hiley, 1993, pp. 147-51). Non-locality was once believed to be a problem but this is no longer the case. As we saw in the previous chapter, any physical theory that agrees with the observed experimental facts of our universe **must** be non-local.

Pilot-wave theory in practice

We cannot reasonably be expected to know Ψ, the wavefunction of the entire universe. In practice we are forced to be satisfied with using the *effective wavefunction* ψ of the tiny subsystem that we are investigating. The methods of estimating ψ are complicated, but in the end wavefunctions from the quantum recipe are used (Dürr, Goldstein & Zanghì, 2013, pp. 37-44).

If the universe is truly deterministic, how do probabilities arise in pilot-wave theory? The answer is that such probabilities are no more than a measure of our ignorance: At any given time t_0, we do not know exactly where the particles under investigation are located. All we know is that their positions are given statistically by the Born Rule which says that the positions of these particles have probability distribution given by $|\psi|^2$. Moreover, we know that no experiment could improve our knowledge.

Experiments are inevitably repeated many times. A single experiment (say a Young's Slits experiment with a single photon) is in effect the selection of a trajectory, taken at random in accordance with the relevant wavefunction ψ. A repetition of this experiment with N photons is, in effect, a random selection of N such trajectories. David Bohm showed that his quantum potential had the wonderful property that, if a "random" (through ignorance) selection of particles satisfied the Born Rule at any given time t_0, then the same rule would be satisfied at all past and future times. This property is called *equivariance*, and a diagram by Travis Norsen (2017, p. 185) is especially instructive as to how it works. In effect, David Bohm has **proven** the Born Rule, which would otherwise be an arbitrary additional axiom of the quantum recipe.

Key examples

Quantum experiments have a different explanation under pilot-wave theory. Here is a brief summary of a few of them.

Recall that, in the **Young's Slits** experiment, there is a beam of identical particles. Because of this, the wavefunction ψ of the beam can be regarded as existing in physical space rather than in configuration space. For the same reason, the quantum potential derived from ψ can be plotted out for different points in space (it is static over time in this instance). This quantum potential has the following shape between the slits and the target (the region on the right-hand side of Figure 4.1): Immediately in front of the pair of slits, the potential has a needle-like peak – this is much narrower than the distance between the slits. A little beyond this, the potential rapidly descends to a plateau, with the exception of narrow, deep ravines, which fan away from the peak, in the direction of the target screen.

Corresponding to this potential, the trajectories of the particles, depending on their exact starting positions as they pass through the slits, can also be plotted. The trajectories fan out because of the narrowness of the slits. On the plateaux, the particles travel in more or less straight lines. Moreover, each time a given particle crosses a ravine in the quantum potential, which it does obliquely, it is accelerated and then decelerated in the $\pm y$ direction (for direction y, see Figure 4.1). This has the effect of kinking each particle's trajectory.

The pattern of trajectories is instructive. They do not cross one another. The trajectories bunch together at locations where the interference pattern is most intense, thus they explain the latter. There is no possibility of directly measuring the location of particles as they pass through the slits without destroying the interference pattern. However, pilot-wave theory enables us to **infer** this location from the place where a particle lands on the target screen. In particular, particles land in the upper half of the target if and only if they passed through the upper hole of the barrier.

When plots of trajectories were first made in 1979 with the help of a computer it revived interest in pilot-wave theory. They may be seen in Bohm & Hiley (1993, pp. 53-4), and in numerous other sources (for instance, search on the Internet for "Bohm Young's slits image").

Pilot-wave theory explains that apparatus – placed so

as to measure directly which slit a particle passes through – significantly changes the quantum potential, thus destroying the interference pattern. In Tim Maudlin's example (with an electron as the particle), the measuring device consists of a proton trapped in a minuscule chamber. This changes the relevant ψ. In particular, the configuration space of ψ increases in dimension, from (x_e, y_e, z_e), to $(x_e, y_e, z_e, x_p, y_p, z_p)$. In consequence, the quantum potential, which is mathematically dependent on ψ, is also radically altered.

For **Spin** experiments, new features of the pilot-wave theory come into play. Imagine a Stern-Gerlach apparatus in which the large magnet is an electromagnet. When the electricity is switched off then, in effect, the magnet does not exist, and neither does its magnetic field. In this situation, spin is irrelevant and so the particles follow trajectories which never overlap. When power to the magnet is switched on, particles having different spin, which previously had identical trajectories, would begin to diverge. This is because the same magnetic field exerts a different force on particles depending on their spin (as explained in subsection ***Physical meaning of spinors*** of chapter 4). Such branching of trajectories never occurs in situations where spin is irrelevant.

A second feature emphasised in spin experiments is that what we **actually** measure, when we measure **any** physical property (be it spin, momentum, energy, or whatever) in reality boils down to measuring **positions** of particles that interact with the apparatus. Thus, although the quantum recipe involves attractive mathematical symmetries – for instance between position and momentum – what is physically important in pilot-wave theory is position.

A final feature is that, in pilot-wave theory, spin is a contextual property – it belongs to the particle only in the context of its being measured by a piece of physical apparatus (in a specific physical orientation). For this reason, a particle cannot possess spin in two different orientations θ_1 and θ_2 simultaneously. Contrast this with a particle's position, which in pilot-wave theory is a property that belongs to it absolutely.

Advantages

In pilot-wave theory there is no artificial division of the world into two parts having different (quantum versus classical) descriptions. Everything that exists is quantum in its nature. A macroscopic object is recognised as such on the grounds that it consists of a large number of particles following very similar trajectories because their positions are entangled. The precise behaviour of a macroscopic object (say a bicycle, or a pointer on a measuring device) can be explained in principle in terms of its constituent particles, as guided by the quantum potential. Such objects are not essentially different from subatomic objects, nor are they subject to different physical laws – even an object as complex as 'an observer' comes within this category.

There is no measurement problem in pilot-wave theory because measurement is not fundamental. One simply recognises that (say) a pointer has moved to a given position, and infers that its particles have moved there together under the guidance of the quantum potential.

Schrödinger's cat dies when the trajectories of its constituent particles become disorganised. If it is confined in a box then we cannot measure directly if or when this circumstance occurs. But, once the box is opened, we can infer the cat's prior status, just as we would if we inhabited a classical world: If the cat is seen to be alive, then it was so all along. If it is dead, then we can infer its time of death, for instance by its degree of rigor mortis. Such indirect inferences are analogous to the inference as to which slit a particle passed through in a Young's Slits experiment, based on the evidence of where it landed.

Pilot-wave theory provides many novel insights into what is going on physically. For example, according to the quantum recipe, some low-energy particles can penetrate a potential energy barrier that, if classical physics were valid, would reflect them all. In 1982 Dewdney and Hiley published a paper showing graphically how a wave-packet approaching the barrier behaves. Only particles at the front of the packet make it through the barrier. Those immediately behind spend a long time in the barrier. Behind them come particles that spend a short time in the barrier. Finally, particles at the rear of the wave packet are reflected even before reaching the barrier. These results are summarised in Bohm and Hiley (1993, pp. 73-8) including diagrams of the quantum potential and trajectories. The (probability distribution of the) time each particle spends in the barrier before being reflected/transmitted is very difficult to discuss – let alone calculate – using Copenhagen methods.

Critiques

Criticisms have been levelled at pilot-wave theory, most with little validity.

First, advocates of pilot-wave theory were accused of wishing to return to classical ideas. However, the law of motion, in which the quantum potential appears, is very different from any classical physical law. Moreover, the non-locality of the universe was first proposed by David Bohm, and this is certainly not a classical idea.

Second, pilot-wave theory was criticised for producing identical results to Copenhagen quantum mechanics. But it was just a historical accident as to which interpretation came first; so why should Copenhagen be preferred on these grounds? Moreover, had pilot-wave theory come first, would anyone wish to swap its clarity and self-consistency for the polar-opposite characteristics of Copenhagen?

Third, particle trajectories were criticised on the grounds that they are "hidden variables" (a term Bohm himself introduced, and later came to regret). As John Bell remarked:

> It is thus from the [particle positions] rather than from ψ, that in this theory we suppose 'observables' to be constructed. ... Thus it would be appropriate to refer to the [particle positions] as 'exposed variables' and to ψ as a 'hidden variable'. It is ironic that the traditional terminology is the reverse of this.
>
> (Brackets denote minor editing, Bell, 1981b, p. 626)

Fourth, particle trajectories are 'metaphysical' because they cannot be measured exactly, even in principle. However, the result of a single run of a high energy physics experiment is often depicted on laboratory computer screens (having a certain resolution) as a network of trajectories. (For instance, search on the Internet for "Higgs boson image".) It is not overly 'metaphysical' to suppose that such particles actually have trajectories that are more accurate than is depicted. For the same reason, it is speculative – but surely not bizarre – to hypothesize that these particles might have exact trajectories.

Fifth, some (including de Broglie when he came to study Bohm's work) have argued that the wavefunction cannot be real because it is defined over configuration space rather than over physical space. I do not think this objection is valid for two reasons. First, philosophers should have absolute freedom (within the bounds of consistency with logic and the known results of physics) for choosing what appears in the ontology. Second, the assumption that only fields defined over physical space (or spacetime) have validity does not come from any logical or physical necessity, but rather from an unquestioned presupposition of physicalism.

The remaining objections are more substantive: Sixth, the wavefunction acts on trajectories but not vice versa. This is an admittedly ugly feature of the ontology. A related feature is that macroscopic objects, because they are spatially entangled, can move to one branch of the wavefunction, leaving the empty branches to continue to evolve according to the Schrödinger equation, but having no significant effect. My opinion is that one can, perhaps reluctantly, simply accept these disagreeable consequences.

Seventh, until recently it has been argued that pilot-wave theory is difficult to make compatible with relativity and cannot explain the creation and annihilation of fundamental particles. Some work has, however, recently been done in this area, see Dürr, Goldstein & Zanghì (2013, Part III).

References

For philosophical discussions of pilot-wave theory, see Lewis (2016), and Maudlin (2019). Both philosophy and physics are given equal weight by Norsen (2017). Physics texts include Bohm (1980), Bohm & Hiley (1993), and Dürr, Goldstein & Zanghì (2013).

Many-worlds

In 1957, Hugh Everett III proposed what is now popularly known as the *many-worlds interpretation* of quantum mechanics. The ontology of this interpretation is that only the wavefunction Ψ of the entire universe exists. This always follows the Schrödinger equation; in particular, it never collapses. In contrast to pilot-wave theory, there are no particles, and hence no trajectories. However, it is still helpful to use these terms when describing experiments.

The theory is best illustrated by an example. Suppose we have a spin-½ particle, which is in the state x-spin up. We measure the state of its spin with a z-oriented Stern-Gerlach apparatus. As described in chapter 4, such a particle will be found to be z-spin up with probability one-half, and z-spin down with the same probability. As was shown in Figure 4.3, the outcome is given by where the particle lands on the target.

Let $|x\uparrow\rangle$ be the initial state of the particle. Let $|\text{Az-init}\rangle$ be the initial state of the Stern-Gerlach apparatus, and let $|\text{Az-up}\rangle$ be the final state of the apparatus if it happens to measure the particle as being z-spin up. Similarly, let $|\text{Az-down}\rangle$ be the final state of the apparatus if it happens to measure the particle as being z-spin down. Lastly, we will use the symbol '→' to indicate the evolution of Ψ over time. According to conventional quantum mechanics, one of two things happens. **Either**:

$$|x\uparrow\rangle\ |\text{Az-init}\rangle \rightarrow |z\uparrow\rangle\ |\text{Az-up}\rangle \qquad \text{with probability } ½$$

Or:

$$|x\uparrow\rangle\ |\text{Az-init}\rangle \rightarrow |z\downarrow\rangle\ |\text{Az-down}\rangle \qquad \text{with probability } ½$$

Everett regarded the observer as part of this system and, defining observer states $|\text{O-init}\rangle$, $|\text{O-up}\rangle$, and $|\text{O-down}\rangle$ in the obvious way, he obtained two expressions:

$$|x\uparrow\rangle\ |\text{Az-init}\rangle\ |\text{O-init}\rangle \rightarrow |z\uparrow\rangle\ |\text{Az-up}\rangle\ |\text{O-up}\rangle$$

$$|x\uparrow\rangle\ |\text{Az-init}\rangle\ |\text{O-init}\rangle \rightarrow |z\downarrow\rangle\ |\text{Az}\downarrow\rangle\ |\text{O-down}\rangle$$

In other words, at the end of the experiment the particle might finish in state $|z\uparrow\rangle$, in which case the apparatus has recorded the particle as being in this state; moreover, the observer's state is one in which he has witnessed the apparatus' record (in the form of memories laid down in his brain). Similarly, if the apparatus had measured the particle as being z-spin down, then the observer would get into a state wherein he had witnessed the apparatus' record of this. It is a core tenet of many-worlds theory that observers are physical entities, and that each observer's brain state gives rise to their mental state.

The conventional view is that the above two expressions are alternative ways in which the wavefunction might collapse.

Everett denies that this ever happens. Instead, he claims that it remains in superposition:

$$|x\uparrow\rangle\,|Az\text{-init}\rangle\,|O\text{-init}\rangle \rightarrow$$
$$\frac{1}{\sqrt{2}}\,|z\uparrow\rangle\,|Az\text{-up}\rangle\,|O\text{-up}\rangle + \frac{1}{\sqrt{2}}\,|z\downarrow\rangle\,|Az\text{-down}\rangle\,|O\text{-down}\rangle$$

In consequence, the initial observer has himself split into a superposition of two observers, $|O\text{-up}\rangle$ and $|O\text{-down}\rangle$. (Other experiments may be such that the observer is split into more than two parts.) Everett introduces the mathematical concept of *relative state* to argue that each of these doppelgangers indeed observes the appropriate state of the measuring device (see Norsen, 2017, p. 277).

We can imagine a more sophisticated S-G apparatus having a needle that points to the words 'up' or 'down', depending on the measurement. Everett writes:

> Why doesn't our observer see a smeared out needle? The answer is quite simple. He behaves just like the apparatus did. When he looks at the needle (interacts) he himself becomes smeared out, but correlated to the apparatus, and hence to the system... [T]he observer himself has split into a number of observers, each of which sees a definite result of the measurement... As an analogy one can imagine an intelligent amoeba with a good memory. As time progresses the amoeba is constantly splitting.
> (Quoted in Norsen, 2017, p. 276)

Recall from chapter 4, that, particularly with macroscopic systems, (the support of) the wavefunction Ψ frequently splits into a number of disjoint, non-interacting branches. In a footnote, Everett comments, "This total lack of effect of one branch on another also implies that no observer will ever be aware of any 'splitting process'" (quoted in Norsen, 2017, p. 280). It is because the state of the universe splits again and again into non-interacting branches that many-worlds received this name – but it would be more accurate to say that there is a single, branching world. In popular works this is sometimes expressed as, "Everything that can happen does happen."

The probability problem

Quantum mechanics contacts reality solely through its theoretical predictions of the probabilities of experimental outcomes, as these are observed by scientists. What probabilities does many-worlds theory predict? Let us imagine that the above experiment is performed three times independently – i.e., using a different particle for each trial rather than repeatedly measuring the same particle. Every time a trial is performed, the wavefunction splits into two branches as shown in Figure 6.1.

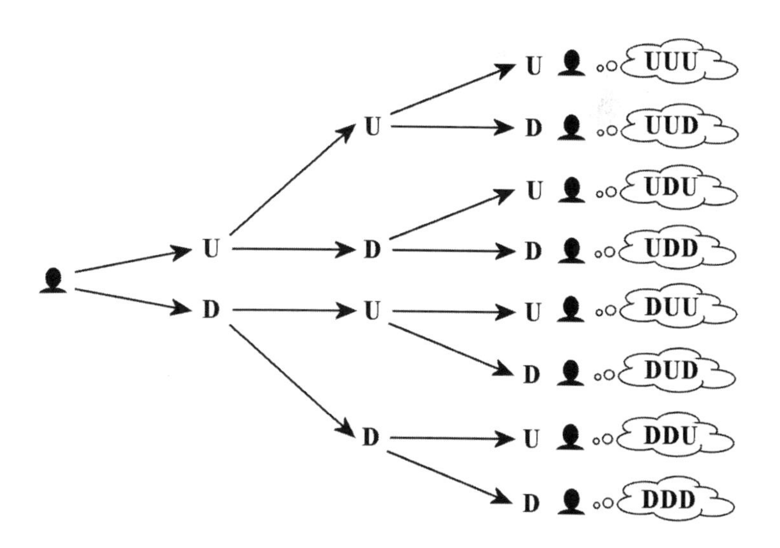

Figure 6.1: Outcomes of three independent repetitions of a spin experiment according to the many-worlds interpretation.

As seen in the Figure, at the conclusion of the three repetitions, one doppelganger will have seen the result **UUU**; four doppelgangers have seen results more ups than downs (in any order); and so on. Figure 6.1 is analogous to the tree-diagram showing possible outcomes of three tosses of a fair coin. In this particular example, many-worlds theory makes correct predictions about the probabilities that outcomes are observed.

But now suppose we repeat the experiment with particles initially oriented in a direction θ away from z in the z-x plane (above we had θ = 90°). As we saw in chapter 4, the experimental evidence is that, for small θ, Prob(**U**) is large and Prob(**D**) is small. But many-worlds theory predicts exactly the same ('fair coin') probabilities as before. This is because each one of the doppelgangers to the right of Figure 6.1 are equally real according to many-worlds theory, and they each observe exactly the same overall result as before – as shown in their thinks bubbles. Thus, with rare exceptions, many-worlds theory contradicts the experimental evidence.

Hugh Everett himself tried to get around this problem by assigning weights to different branches of the wavefunction, but he gave no adequate motivation for doing this. Indeed, assigning such weights goes against the very crux of his theory, which is that, were you to perform such an experiment, all of your unseen doppelgangers would be no less real than you are. Subsequently, many different attempts have been made to solve this problem. Some authors, rather than focussing on what is the case, instead discuss what an observer might 'rationally' believe the outcome is likely to be, or how much we should 'care' about each outcome. But such considerations are unsatisfactory and irrelevant because **according to the theory itself** every outcome occurs.

In actual practice, many-worlds theorists finagle some argument that gives them an excuse to use the Born Rule. On these grounds, they claim that many-worlds gives the same result as the quantum recipe.

Advantages and disadvantages

Despite this, the many-worlds interpretation is surprisingly popular. For cosmologists, many-worlds has the advantage that it allows discussion of the entire universe. Another benefit is that observers are explicitly a natural part of the universe, following the same laws as any other entity.

A frequently-made criticism of many-worlds is that it is metaphysically extravagant. The proliferation of branches is prodigious and has gone on at every instant, at least since the Big Bang. This criticism is not – in itself – valid. Such profligacy would be of no importance, **but only if** the theory could be shown to be empirically adequate.

Among its disadvantages, many-worlds lacks clarity because it lacks any local beables: Ψ is a function of (multidimensional) configuration space rather than (3+1 dimensional) physical space: it is unclear whether or not items such as tables or tigers can be represented adequately as structures in configuration space. But the bottom line is that, even considered in isolation, the probability problem is fatal.

References

Philosophical discussions are given by Lewis (2016, see "many-worlds" indexed on p. 205), and by Maudlin (2019, chapter 6). Both philosophy and physics are given equal weight by Norsen (2017, chapter 10).

Ghirardi, Rimini and Weber

The Ghirardi, Rimini and Weber (GRW) interpretation is one in which the wavefunction Ψ for the universe is a real entity and undergoes spontaneous collapses **according to specific physical laws**. These laws have been carefully chosen so as to give results indistinguishable from quantum mechanics for all experiments that have been carried out to date. It is conceivable that, in the future, some very delicate experiment will be able to establish whether or not GRW is true; but, for over thirty years, this has not happened. The possibility of refutation is not a weakness of GRW, because falsifiability is a defining characteristic of scientific theories.

The original GRW paper (1986) is highly technical, but it begins with a detailed and accessible introduction. Its title is important: "Unified dynamics for microscopic and macroscopic systems". In other words, under the GRW interpretation, in contrast to Copenhagen, entities of any size obey exactly the same physical laws. The aim of the paper is also crucial; it is to give an interpretation in which well-defined trajectories emerge for macroscopic objects (GRW, 1986, p. 471). Each spontaneous collapse is a localisation process that "is formally identical to an approximate position measurement" (p. 471). A thorough account of GRW theory, intended for non-specialists, was later written by Giancarlo Ghirardi himself (2007; throughout, but especially chapters 14-17).

One can introduce the GRW interpretation as having both waves and particles, and I will do so here. But, strictly speaking, only the wavefunction of the universe is essential. (Particles and macroscopic objects are implicit in the wavefunction.)

Formalism

Here, with some minor notational changes, I will make use of John Bell's article, "Are there quantum jumps?" (1987, pp. 3-7). The wavefunction may be expressed as:

$$\Psi(\mathbf{q}_1, \mathbf{q}_2, \ldots, \mathbf{q}_N; t)$$

This notation indicates that: there are N particles; $\mathbf{q}_k = (x_k, y_k, z_k)$ are the positional coordinates of the kth particle; and t is time. In line with what was said above, whenever the 'kth particle' is mentioned here, strictly speaking, this is merely a reference to the kth triplet of arguments of Ψ.

Usually, the wavefunction will follow the Schrödinger equation. But at random times, and for particles (more strictly, subscripts k) chosen at random, the wavefunction jumps or *localises* to a 'reduced' or 'collapsed' wavefunction:

$$\Psi' = \mathbf{J}(\mathbf{q}^0 - \mathbf{q}_k)\ \Psi(\mathbf{q}_1, \mathbf{q}_2, \ldots, \mathbf{q}_N; t)\ /\ R_k(\mathbf{q}^0)$$

This takes some unpicking: Ψ' (psi-prime) is the new wavefunction after the collapse. **J** is the jump function which localises the kth particle, in such a way that it must end up very near position $\mathbf{q}^0$. $R_k(\mathbf{q}^0)$ is just a function that is defined so as to normalise Ψ', as described in chapter 4.

Answers to questions

Which particle will be localised? Particles are localised by Nature at random – with equal probability for each of them.

How often is each particle localised? This also happens at random, but on average about once in 10^{15} seconds, which is about once in 100 million years. GRW take this to be a new constant of nature, which Bell calls τ. Localisations obey the statistics of the Poisson distribution.

Old-fashioned filament lightbulbs also obey this distribution – they last an average of (say) three years before burning out. But suppose you have one that has already lasted for ten years. How much longer would you expect it to survive? The surprising answer is another three years on average. The time that has already passed has made absolutely no difference to the expected lifetime. This is a remarkable, and defining, characteristic of the Poisson distribution. It follows that, if a particle has not localised for a vast period of time, probabilities about when it will do this in the future are entirely unaltered.

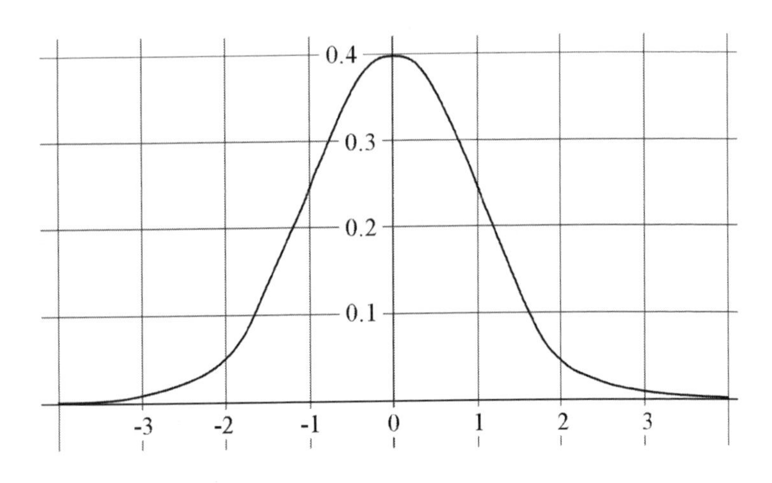

Figure 6.2: The Standard Normal distribution, with mean 0.0, and standard deviation 1.0. The area under the curve is 1.0.

What is the Jump function **J**? GRW say that this could be any smoothly-changing function so long as it is narrow. They suggest a (three-dimensional) Normal distribution function for **J**, whose *width* (more correctly, standard deviation) is σ. You may recognise the (one-dimensional) Normal distribution as having the shape of a smoothly contoured hill, as shown in Figure 6.2. GRW suggest that a suitable width is $\sigma = 10^{-5}$ cm, and that this is another new constant of nature.

In what manner is the localisation position $\mathbf{q}^0$ *chosen?* The **approximate** answer is that Nature must choose $\mathbf{q}^0$ according to the statistics of the Born Rule. That is to say, $\mathbf{q}^0$ must be chosen in such a manner that the probability of its being in region R is given by the squared amplitude of the wavefunction integrated over that region.

The **exact** rule for choosing $\mathbf{q}^0$ is slightly more complicated, and Bell gives it in his equations (6) & (7) (1987, p. 4). The complication is related to the fact that **J** has a non-zero width. In many cases, the Born Rule above is an excellent approximation to the exact rule, and so the latter will not be explained here.

Ontology

The ontology of GRW consists for the most part in the wavefunction Ψ of the universe. Usually, Ψ follows the Schrödinger equation, but occasionally it undergoes spontaneous collapse.

The Jump function **J** can be regarded as part of the ontology. Its width, σ, is a constant of nature. The average time, τ, it takes for a given particle to localise is another.

The final part of the ontology is the local beables. These give empirical content to the theory. With GRW, there are two major proposals, "flashes" and "mass density". These alternatives will be discussed next.

GRWf

In "flashy GRW", or GRWf for short, the local beables are *flashes*. Suppose that Ψ collapses to Ψ' at time t^0, as described by the formula for Ψ' above. Then the corresponding flash occurs at spacetime point $(\mathbf{q}^0, t^0)$. Bell, who first made this proposal, said of flashes:

> These are the mathematical counterparts in the theory to real events at definite places and times in the real world ([...] as distinct from the 'observables' of other formulations of quantum mechanics for which we have no use here). A piece of matter is then a galaxy of such events.
> (Bell, 1987, p. 7)

Each flash, of itself, provides no information as to which particle k was involved – it does no more than exist at a specific location. Under certain circumstances, however, a set of flashes may be so arranged in spacetime as to display a well-defined trajectory.

Flashes, whilst numerous at the everyday scale, are extremely sparse at the microscopic scale. Tim Maudlin points out that the human body, from its electrons alone, causes about ten trillion flashes per second. But, because the human body contains many cells this amounts to only a few flashes per second per cell (Maudlin, 2019, p. 113). (Of course, in the very different situation in which a human cell is being examined using a microscope, there are many more flashes that are caused by the light that illuminates the cell before entering the microscope.)

GRWm

In "massy GRW", or GRWm for short, the local beable is a *mass density* for each particle. To give slightly more detail: although Ψ is defined over configuration space, we want to define mass density ***MD*** over each point **q** = (x, y, z) in physical space in some natural way. This is done using the inner product:

$$MD(\mathbf{q}) = \langle \Psi \mid M(\mathbf{q})(\Psi) \rangle \text{ standardly written as } \langle \Psi \mid M(\mathbf{q}) \mid \Psi \rangle$$

Where, for each **q**, M(**q**) is the **mass density linear operator** that acts on Ψ. M(**q**) is defined by some highly complex formula, not presented here – ultimately it is derived from the energy operator H, without which the Schrödinger equation lacks specific content.

How is mass density *MD*(**q**) to be understood?

- ***Standard*** quantum mechanics interprets the above expression for *MD*(**q**) as giving the ***expected (or average)*** value of mass density of matter at **q**, ***when this is measured***

In contrast,

- ***Massy GRW*** interprets *MD*(**q**) as ***being the actual*** mass density of matter at **q**, ***irrespective of any "measurement"***

The emphasis above indicates the points of contrast. See Lewis (2016, fn. 4 on p. 192).

According to GRWm, the wavefunction Ψ occasionally undergoes a spasm such that the mass density of particle k becomes localised. Although it is non-standard terminology, for convenience when discussing GRWm, or GRW in general, I will sometimes still refer to the centre of localisation of mass density ($\mathbf{q}^0$, t^0) as a "flash".

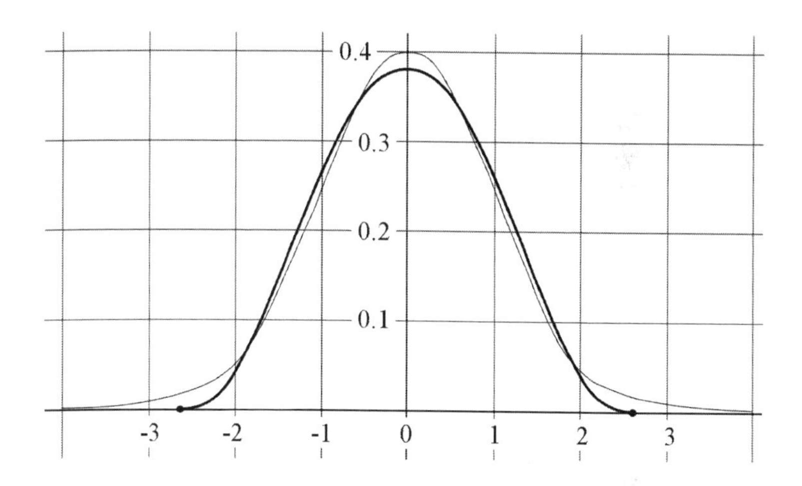

Figure 6.3: Shown in **bold** *is the tailless cosine function T_r with r = 1.2. It is less than the Standard Normal function (shown faint) when $|x| < 0.5$, and also when $|x| > 2$. Elsewhere, T_r is greater. The region where T_r is non-zero is $(-\pi/1.2, \pi/1.2) \simeq (-2.6, 2.6)$.*

A well-known problem with GRWm is that, even after collapse, a minuscule portion of the mass distribution of the collapsed particle still extends off to infinity. This, however, is an artefact of the idealised manner in which the Normal distribution has been defined mathematically. (For instance, heights are approximately Normally distributed, but no one believes that an extremely rare person will grow be ten metres tall!) An idea of my own is that we can replace the Normal jump function with a tailless (truncated) cosine function of the form:

$$T_r(x) = \frac{r}{2\pi}(1 + \cos(rx)) \quad \text{for } x \text{ in } \left(-\frac{\pi}{r}, \frac{\pi}{r}\right), \text{ and 0 elsewhere}$$

The area under $T_r(x)$ is exactly 1 for any r. When we take r = 1.2, then $T_{1.2}(x)$ gives an excellent approximation to the Standard Normal distribution N, as can be seen in Figure 6.3.

This function can be generalised to three dimensions in the obvious way. Making this alteration in the jump function, from **J** = N to **J** = $T_{1.2}$, makes no practical difference other than to eliminate the unwanted and unnatural tails.

The new physical constants τ and σ

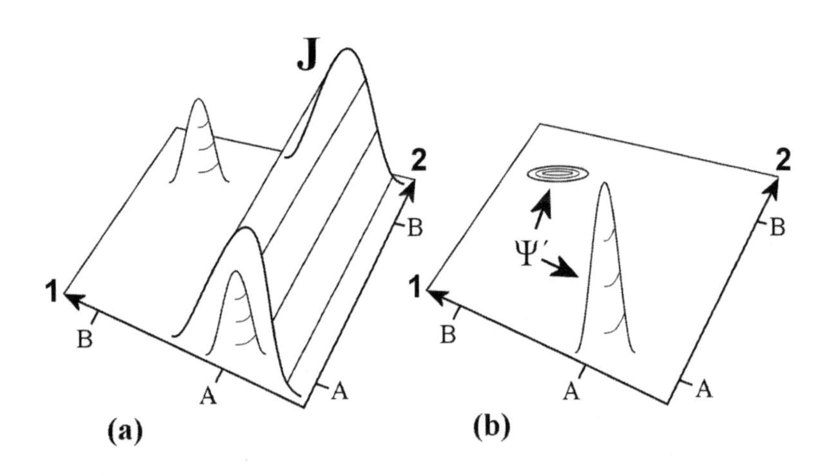

*Figure 6.4: Spontaneous localisation according to the GRW interpretation. In **(a)** wavefunction Ψ happens to consist of two equal peaks. Particle 1 is about to undergo a localisation according to jump function J. Sketch **(b)** shows the new wavefunction Ψ′ after localisation.*

GRW is specifically designed to give predictions that are indistinguishable from this of the quantum recipe for all known experiments to date. The physical constants τ and σ were chosen so as to ensure this. We will examine each in turn.

The physical constant τ: GRW chose τ to be very large (about 10^8 years) in order that, in physics experiments involving a very few particles, or a beam of such particles, or a few molecules, localisations have an essentially zero probability of occurring.

However, macroscopic objects, such as pointers, measuring devices, and cats, consist of a multitude of entangled particles. When any single particle that makes up such an object undergoes a localisation, then this causes the wavefunction of all of the particles of that entity to collapse. Figure 6.4(a) sketches, in greatly simplified form, the configuration space of a system composed of two entangled particles, each constrained to a single dimension x. The entanglement is a superposition such that either both particles are located near A, or both are located near B. With the obvious notation, in Figure 6.4(a) we have that the relevant portion of the universe's wavefunction is given by:

$$\Psi = \frac{1}{\sqrt{2}} |\mathbf{1}_A\rangle |\mathbf{2}_A\rangle - \frac{1}{\sqrt{2}} |\mathbf{1}_B\rangle |\mathbf{2}_B\rangle$$

(**Aside:** the reason there is a minus rather than a plus sign in the above expression is that we are dealing with constituents of matter (fermions). This will be explained in chapter 10. The sign is not relevant to the following argument.)

The figure also shows the jump function **J**, which localises particle **1** near A. You can see that **J** has an identical profile at all positions in the direction of axis **2**.

Figure 6.4(b) shows the resultant wavefunction Ψ′ after localisation. For the most part, both particles are now located near A. The other peak has all but disappeared, being reduced to a 'pancake' of infinitesimal height. If my suggestion of a tailless jump function is adopted, then the pancake and the superposition vanish entirely.

Now consider a macroscopic object, consisting of a large

number of entangled particles, **1**...**N**, which is in a superposition of being in locations A or B. Suppose particle **1** localises near A. The result is that all of the particles belonging to the object also localise near A. This can be represented in Figure 6.4(a) & (b) by replacing the axis label "**2**" with "**2...N**". Because such objects generate billions or trillions of localisations per second, they cannot remain in a superposition of being at separated locations for more than the briefest instant. Instrument pointers never become smeared out, and Schrödinger's cat is in an extremely well- (not perfectly) defined state, even when incarcerated in the locked box.

The physical constant σ: Whilst $\sigma = 10^{-5}$ cm is small on everyday scales, at first sight this width is surprisingly large. After all, according to the quantum recipe, the wavefunction effectively collapses so as to give a definite value for the position that has been measured: this value being the one that was actually observed. Why can't we take σ to be very close to 0? The answer is that, if a particle localised within an atom, it would deposit a large amount of energy there, and likely ionise or even destroy the atom. This is not seen to happen. The jump function **J** is therefore much broader, leaving the position of the particle vague even after localisation. The width $\sigma = 10^{-5}$ cm is about 200 times the size of a hydrogen atom. This means that the particle is far from certain to be within the ambit of even the largest molecule, even when the flash occurs at the latter's centre. Moreover, broader localisations are, by their nature, less energetic. See Maudlin (2019, p. 104).

Stability of atoms

Suppose an electron is in an orbital within an atom, and has wavefunction Ψ. This circumstance means that its position is very narrowly fixed. Suppose the electron itself were to undergo a localisation. The approximate Born Rule means that the flash (i.e., the centre of **J**) is overwhelmingly likely to occur where the electron is. The jump function **J** is hundreds of times wider than the atom, and so it has an almost constant value, call it j, over the support of Ψ. This means that $\Psi' = \mathbf{J}\Psi = j\Psi$. When we renormalise, we get $\Psi' = \Psi$. At most, a localisation negligibly "tickles" an electron that is bound within an atom. See Maudlin (2019, p. 105).

If we had found that localisations had a significant effect on electrons in orbitals, then this would have implied that all atoms are slightly unstable. The effect would be readily observable because of the vast number of atoms, within even a small object. But our experience is that (with a few specific exceptions) atoms are remarkably stable over vast timescales. The theoretical finding, that GRW correctly predicts the stability of atoms, is reassuring.

Spin

Spin presents no problems for GRW. Localisations **J** happen to position variables only: spins themselves are never localised. A z-spin up particle passing through a z-oriented Stern-Gerlach apparatus (Figure 4.3) will always be deflected up:

$$|z\uparrow\rangle\,|\text{undeflected}\rangle \rightarrow |z\uparrow\rangle\,|\text{deflected up}\rangle$$

Likewise, a z-spin down particle passing through this S-G apparatus will always be deflected down:

$$|z\downarrow\rangle\,|\text{undeflected}\rangle \rightarrow |z\downarrow\rangle\,|\text{deflected down}\rangle$$

A particle spinning in an arbitrary direction, $c|z\uparrow\rangle + d|z\downarrow\rangle$, where c and d are complex numbers, can be regarded as a superposition of the above cases:

$$(c\,|z\uparrow\rangle + d\,|z\downarrow\rangle)\,|\text{undeflected}\rangle \rightarrow c\,|z\uparrow\rangle\,|\text{deflected up}\rangle + d\,|z\downarrow\rangle\,|\text{deflected down}\rangle$$

As shown in chapter 4, a beam of many randomly oriented particles will be split into two equal portions:

$$|\text{undeflected beam}\rangle \rightarrow$$
$$(1/\sqrt{2})\ |\text{deflected up}\rangle + (1/\sqrt{2})\ |\text{deflected down}\rangle$$

As the superposed deflected beams approach the target, they are separated by at least 1 mm, far more than the width σ of the localisation function **J**. When **J** occurs, because of the approximate Born Rule, it will be in one of the areas "z-spin up", "z-spin down" on the target, which are labelled in Figure 4.3.

(Followers of GRW need to explain how a localisation event – corresponding to an energetic particle within the beam hitting the target at an appropriate place – gives rise to a mark on the target at that spot. Moreover, this explanation should be given solely in terms of the ontology of GRW. This will be explained in chapter 10.)

Bell's-inequality experiments

GRW can explain Bell's-inequality experiments. Suppose a pair of entangled spin-½ particles, **1** and **2**, are emitted from a source, and travel to remote locations where z-oriented S-G apparatuses are set up. The entanglement is such that if particle **1** is measured to be z-spin up, then the other will be measured to be z-spin down, and vice versa. Assume that the apparatus to measure particle **1** happens to be slightly closer to the source. Then this apparatus will be the first one to trigger a localisation. The situation is as depicted in Figure 6.5.

The jump **J** could have occurred anywhere within the support of the combined beam of particle **1**. The two sub-beams of particle **1** are centred at "**1** up" and "**1** down" in Figure 6.5(a). Because of the approximate Born Rule, **J** might occur near either of these latter positions. It so happens that, in the Figure, the localisation is at "**1** down", i.e., in the "z-spin down" component of the particle

1 beam. (But it is equally likely that **J** would occur near "**1** up".)

In Figure 6.5(b), the black dot • shows where Particle **1** flashed on its target. Because the wavefunction has collapsed to Ψ', particle **2** must flash at the white dot **o** a moment later. In this instance, particle **1** is measured as being z-spin down, and particle **2** as z-spin up.

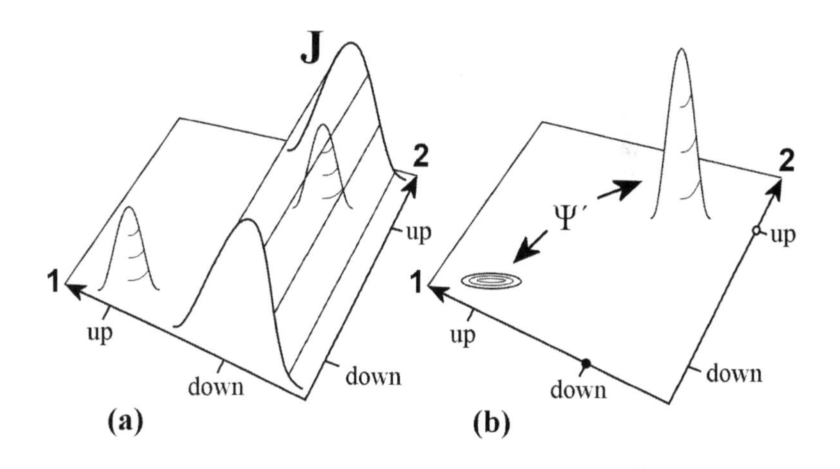

*Figure 6.5: A Bell's-inequality experiment according to the GRW interpretation. Particles **1** and **2** are distant from one another. In succession, they encounter the targets of their respective S-G apparatuses. Positions on these targets have been labelled "up" and "down".*

It should be fairly clear that GRW reproduces the correct predictions of the quantum recipe. This is so, even if the apparatuses are oriented at different angles. It can be shown that similar arguments go through for other variants of Bell's-inequality experiments, which may make use of different properties of particles, and different measuring devices.

References for GRW

Philosophical discussions are given by Lewis (2016, see "collapse" and "GRW" in the index), and by Maudlin (2019, chapter 4). Philosophy and physics are given equal weight by Norsen (2017, chapter 9). Bell was an avid supporter of GRW; in addition to the references already given, see (Bell, 1990, pp. 39-40). Ghirardi (2007) is an authoritative account intended for non-specialists.

Choosing an interpretation

My intention in Part III is to show that the mind-body problem is not a scientific question – instead, it is philosophical in its character. Mind is not part of physics. Despite this, the solution to the problem must be a rational one; furthermore, it has to be consistent with present-day science. Given this approach, my first key criterion for a suitable interpretation of quantum mechanics is that it must give an objective account of the physical world, with no mention of mind.

Copenhagen fails this criterion because it brings the concept of the 'observer' into physics. Attempts made by some members of this school, to give a more objective account of the universe, lack clarity. This interpretation is therefore unsuitable for my purposes. Many-worlds attempts to bring human minds (percepts and memories) into physics, as sketched in Figure 6.1. It therefore also fails my criterion. Moreover, as argued earlier, this theory is empirically inadequate.

Pilot-wave theory has the advantage that, consistent with my criterion, it treats 'observers' like any other physical objects. A major objection for my purposes is that it is strictly deterministic – once the initial wavefunction and the trajectories of all the particles are given at the dawn of time, the universe thereafter goes on its unswerving way. Appending consciousness to such a model could allow it no causal role: we would be helpless spectators to every event in our lives. While the general consensus asserts that such freedom as we possess is compatible with determinism, I find such arguments unconvincing. In particular this position is in stark contradiction to the evolutionary advantages that consciousness seems to bring to complex creatures. Mind, if it plays no causal role, cannot contribute to an animal's biological fitness. Despite its attractions for many other purposes, I therefore reject pilot-wave theory as a candidate in my attempt to address the mind-body problem in Part III of this book.

My chosen interpretation for Part III is GRW. This is because it has two essential features: First, consistent with my key criterion, it gives a clear, objective account of the universe. Second, according to GRW the universe is stochastic in its character. The latter is a second key criterion because I wish to give a rational account of free will. The consensus view is that physical randomness is useless in giving humans room to exercise free will. Schrödinger makes this case forcefully – he clearly articulates all of the essential arguments of the compatibilist position. In chapter 9, I will develop a theory of free will, and provide a detailed response to Schrödinger. There are several other interpretations of quantum mechanics, and any that possesses the two key criteria would be suitable for my purposes.

Between GRWm and GRWf, I have a slight preference for the flash ontology. This is because the massy ontology suggests that physics deals with some objective 'stuff'. In contrast flashes, although they occur objectively at specific locations in space and time, are suggestive of things that are experienced. As John Bell puts it, referring to flashes as "events":

> As a schematic psychophysical parallelism we can suppose that our personal experience is more or less directly of events in particular pieces of matter, our brains, which events are in turn correlated with events in our bodies as a whole, and they in turn with events in the outer world.
> (Bell, 1987, pp. 7-8)

Part III

Pan-idealism – a solution

Chapter 7

Physics and concrete reality

Here the case will be made that the concept of **concrete reality** has not been analysed with sufficient clarity until recently. This has led to difficulties, both for physicalists, and for the study of the mind-body problem. Many physicalists still mistakenly believe that concrete reality is a scientific concept, and that everyone has adequate intuitions as to what this term means.

In contrast, my aim is to show that concrete reality (also known as objective existence) is a philosophical concept – outside the scope of science. It is true that concrete existence cannot be a **property** but, having accepted this, what is it? I do not believe that one can try to define concrete existence; instead, one should attempt to characterise it in a clear, consistent, and acceptable manner. In the course of attempting to refine our intuitions, it is inevitable that our initial pre-theoretic ideas will have to be modified.

By the end of the chapter, I hope to have shown that everything that is concretely real **must** be characterised by being mind-like in its intrinsic nature. It is impossible, even on the deepest reflection, to form any meaningful notion of an entity – in this or in any other universe – that is concretely real, but which totally lacks mind.

Science and the universe

What kind of picture does science give of the universe and its contents? The answer is that of a mathematical model. For example: The magnet within a Stern-Gerlach apparatus is a lump of iron of a certain shape. It consists of crystals that are cubical lattices of iron atoms. The atoms themselves have subatomic structures. Electrons move more or less freely among these atoms. There is a magnetic field between the poles of the magnet. All of the words in this written description have a mathematical counterpart. If we want even more detail, then this will boil down to more detailed mathematics. According to quantum mechanics, electrons are not what we would naively regard as particles, and neither are they waves. Despite this, they have a full (at least in terms of science) description given by the mathematics of their wavefunction.

In classical physics, the adequacy of mathematical models in giving an objective account of the universe and its contents is unquestioned. In quantum mechanics, the Copenhagen interpretation denies that we can give an objective account. There are many reasons, which were given in previous chapters, to reject Copenhagen. Better interpretations perform essentially the same calculations as those of the quantum recipe. Further, they make exactly the same predictions as to experimental results. (In the case of GRW, the predictions are merely near-enough the same; i.e., to the accuracy of all experiments to-date.) Each of these more realistic interpretations provides its own distinctive, mathematical model as to what is actually going on in the universe.

The fact that science pictures the world in wholly mathematical terms has led to two types of *structuralist* philosophical positions: *Epistemic structural realism* (ESR) asserts that **all we can know about** the universe is its mathematical structure. This idea dates back to Kant (1781). More extreme is *ontological structural realism* (OSR) which claims that **all there is to** the universe is its mathematical structure: there is nothing else. This has been popular since the 1930s to the present.

Empirical reality versus concrete reality

An entity or property is *empirically real* **if it can be observed or measured** by a human (or perhaps other) intelligence. Empirical reality is a scientific notion. In contrast, an entity or property is *concretely real* (or just *real* for short) **if it exists independently of**

being measured or observed. This is not a scientific notion. As Berkeley was first to point out, this concept of concrete existence independent of measurement or observation is utterly obscure. The clear tension between these two definitions is highlighted by the words in bold above.

Although it is not a scientific concept, most scientists are passionate about reality. Einstein, for example, had many conversations with Abraham Pais, an adherent of the Copenhagen orthodoxy, who reported:

> The main topic of discussion was quantum physics, however. Einstein never ceased to ponder the meaning of the quantum theory. Time and time again the argument would turn to quantum mechanics and its interpretation. [...] We often discussed his notions on objective reality. I recall that during one walk Einstein suddenly stopped, turned to me and asked whether I really believed that the moon exists only when I look at it. The rest of this walk was devoted to a discussion of what a physicist should mean by the term "to exist."
> (Pais, 1979, p. 907)

In contrast, David Mermin once went so far as to assert – as a preliminary to discussing Bell-inequality experiments:

> The questions with which Einstein attacked the quantum theory do have answers; but they are not the answers Einstein expected them to have. **We now know that the moon is demonstrably not there when nobody looks.**
> (Emphasis added, Mermin, 1990, p. 81)

Mermin's first sentence is correct. The second sentence is hyperbolic, untrue and, as Einstein's words suggest, absurd. It implies that the universe does not exist – an unacceptable philosophical position called *solipsism*. Mermin's book nowhere contains a reasoned argument that would justify his second assertion.

The substantive reality of concrete reality

In his excellent 2019 book, Tim Maudlin puts the case that disputes about concrete reality are no more than differences in attitudes that scientists might have to their theories (pp. xi-xiii). He cites the publication of Copernicus' *De Revolutionibus*, in which the author argued that every planet, in truth, orbits the Sun; and contrasted this with Osiander's preface, which argued that the theory should merely be regarded as a calculating device, enabling astronomers to predict the positions of planets, but without making any truth claims. This preface was written anonymously, and without Copernicus' permission.

Is this really a matter of attitude, akin to whether or not one likes Marmite? Or is this disagreement **substantive** – one for which it is appropriate (like Einstein) to make a serious, and rationally based commitment to a particular side?

My view is that concrete reality is a substantive (although not scientific) concept: Some things are concretely real, and others are not. Some putative entities, which we might prima facie regard to be possible examples having concrete reality, cannot be so. This is because, upon reflection, there is no possibility of describing what they supposedly are in a clear and coherent manner. By the end of the chapter I hope to provide rational and lucid criteria **characterising** concrete reality, so as to meet almost all of our intuitions about the concept, but with the exception just mentioned.

All scientific claims must in principle be empirically falsifiable, and this is why the concept of concrete reality cannot be a scientific one.

ESR leads towards irrealism

The difficulties of interpreting quantum mechanics have led many physicists towards ESR, and towards the whole debate about realism versus irrealism of chapter 6. My view is that denying the reality of the Moon when no one is looking, or expressing indifference as to its reality, is no small concession for a physicist to make. It is certainly premature when realistic interpretations of quantum mechanics are to be had. John Bell was scathing about this irrealist tendency. He wrote, with regard to experimental science:

> The aim remains: to understand the world. To restrict quantum mechanics to be exclusively about piddling laboratory operations is to betray the great enterprise. A serious formulation will not exclude the big world outside the laboratory.
> (Bell, 1990, p. 34)

Thin realism

Let *thin realism* be a synonym for OSR: the mathematical description of the universe is all that there is to concrete reality. Although this position may seem attractive, especially to scientists who are strong reductionists, it has a major problem. **Thin realism cannot distinguish between possible and actual worlds.** In particular, it cannot say what it is about the actual universe that distinguishes it from possible, but false, mathematical models.

To give an example, my toothbrush is in my bathroom, and has a certain definite number of bristles. There are plausible mathematical models of the room, which are consistent with modern physics, and which give the correct answer to this question. Likewise there are other, similar models that give an incorrect answer. (Throughout this example, to avoid the complications of quantum mechanics, I'm considering only macroscopic objects and their macroscopic properties – such as toothbrushes and their number of bristles.)

How can I determine which of these mathematical models is

the correct one? It is true that I could go into the room and count the bristles, thus determining this particular fact empirically. But the toothbrush has numerous physical properties – far too many to determine empirically. In any event, having to make an empirical measurement for each concretely existing property goes against the spirit of OSR, which asserts that entities and properties concretely exist independent of me or anyone else measuring them.

The problem is already intractable even when describing my own toothbrush. If we wish to discuss my bathroom as a whole – let alone the entire universe – the difficulties explode without limit.

Tegmark's ultra-thin realism

Physicist Max Tegmark is a thin realist who has seen the force of the above problem better than anyone. He holds that mathematical existence and physical existence are the same concept. We, and the world we live in, are nothing other than an (incredibly complex) formal mathematical system. There is no intrinsic character to any physical object, and there is no extra fact that distinguishes our instantiated world from its mathematical model: "instantiation" is thus an empty concept for him (Tegmark, 2003, section IVB; Hut *et al.*, 2006, section IIA). His reasons for adopting this position are: the extraordinary success of physical theories that describe the world in mathematical terms; and the lack of any scientific evidence for intrinsic, non-mathematical properties in the external world (Hut *et al.*, 2006, pp. 2-3, 12).

Tegmark's conception of consciousness is that there are (mathematically defined) self-aware substructures in a universe. He goes further and proposes that it might be the case that **all** formal mathematical objects have concrete (he says "physical") existence:

> Now suppose that our physical world really is a mathematical structure, and that you are a self-aware substructure within it. In other words, this particular mathematical structure enjoys not only mathematical existence, but physical existence as well. What about [all other mathematical structures]? Do they too enjoy physical existence? If not, there would be a fundamental, unexplained ontological asymmetry built into the very heart of reality, splitting mathematical structures into two classes: those with and without physical existence. As a way out of this philosophical conundrum, I have suggested [...] that complete mathematical democracy holds: that mathematical and physical existence are equivalent, so that all mathematical structures exist physically as well.
> (Tegmark, 2003, p. 14)

Tegmark thus proposes an amazingly wide concept of existence. Not only does the world in which I live exist, so do all of the possible worlds in which my bathroom has different physical contents. To take ever more extreme differences: there exist universes that are identical to our own, with the exception of having different initial conditions; there exist universes where the laws are identical, but in which the constants of nature (h, c, e and so on) are different from our own universe; and finally, there are worlds in which the laws of physics are totally different from our universe. All of these countless other worlds are no less real than our own. Tegmark's concept of concrete existence is thus far more profligate and implausible, even than that given by the many-worlds interpretation of quantum mechanics.

Tegmark's account of consciousness is meagre (as can be seen from his discussion above): it amounts to no more than insisting that thin physicalism must be true.

Thick realism

Let *thick realism* be the position that there is 'something more' to concrete reality than its mathematical description. What can this something more be? Whatever it is, it certainly cannot be part of science. This calls into question as to whether anything meaningful can be said about it at all – some critics have dubbed it 'secret sauce'. Before turning to what I hope to show (by the end of the book) is a successful account of thick realism, I will describe some less successful ones:

Concrete reality as noumenon

As long ago as 1781, the philosopher Immanuel Kant affirmed that there was something else, which he called the *noumenon* of each concretely real entity. But he went on to assert that the noumenon of any entity is entirely unknowable – even in principle. The complete mysteriousness of the noumenon is an absolute impediment to coming to grips with it. For this reason, Kant's later followers did away with the noumenon.

Matter as the permanent possibility of being perceived

John Stuart Mill came up with this idea in his 1865 book *An Examination of Sir William Hamilton's Philosophy*, or *Examination* for short. My source here is an entry in the *Stanford Encyclopedia of Philosophy* (Macleod, 2016). Mill wrote:

> Matter, then, may be defined, a Permanent Possibility of Sensation [i.e., "of being perceived"]. If I am asked, whether I believe in matter, I ask whether the questioner accepts this definition of it. If he does, I believe in matter: and so do all Berkeleians. In any other sense than this, I do not.
> (Expression in brackets added, *Examination*, IX: p. 183)

In other words, my toothbrush in the next room is not being perceived at present. However, because it has a Permanent Possibility of being Perceived (call this *PPobP* for short), I can be confident I will see it on re-entering the bathroom.

Mill was trying to build a philosophical system, having many similarities to Berkeley's idealism, but without the concept of God. In Berkeley's system, God played an essential role in ensuring that human minds had consistent experiences. For Mill on the other hand, matter itself – in virtue of its PPobP – ensures that the successive experiences of a given human individual are consistent; moreover, the same consistency also applies to the experiences of different individuals. Mill's theory is ontologically grounded in mind, but his proposed PPobP is a feature of matter that is not, of itself, mind-like.

Mill always insisted that mind is a natural part of the universe, but he never specified how to reconcile the stark contrast between human mind, and mindless matter possessing PPobP. For this reason, his theory steers dangerously close to Cartesian substance dualism. Another difficulty is that it is hard to understand what PPobP amounts to as a feature of matter.

Causal flux

This is a picture of reality as consisting solely of relationships. Relationships are not between 'things', because there are no things as such in the universe. Instead, relationships are connected to other relationships: the world is nothing other than a complicated network of such relationships. These relationships are considered to have some substantive causal or law-like power or effect. Chalmers describes this briefly, "One might be attracted to the view of the world as pure causal flux, with no further properties for the causation to relate, but this would lead to a strangely insubstantial view of the world" (1996, p. 153). In a footnote (p. 375), Chalmers attributes this position to Shoemaker in 1980, and goes on to critique it.

Apart from its insubstantiality, the causal powers are obscure.

Intrinsic nature

The kind of thick realism that I believe will prove fruitful is that **mind, broadly construed, is the intrinsic nature of all reality**. Call this the *intrinsic nature* position. Throughout this book, anything that has experiences is said to have *mind* – it need not have any cognitive powers. The advantage of this position is that the intrinsic nature of the world is knowable to us, at least in part. We are a natural part of the world, and we have some – admittedly limited, but certainly not zero – insight into our own experiences.

This position implies that mind is present **everywhere** in the universe. The theory that all matter has mental (or mind-like) properties is called *panpsychism*. Clearly, all variants of the intrinsic nature position are types of panpsychism; but there are other versions of panpsychism that rely on concepts distinct from that of intrinsic nature. (See, for example, Russell's neutral monism, discussed below.)

History

In a basic form, the theory that the intrinsic nature of matter is mind has ancient roots. Aristotle, in his book *De Anima* ("On the soul", meaning "Concerning that which animates"), reports that **Thales of Miletus**, who lived ~620 to ~540 BCE, claimed that "everything is full of gods." Thales attributed soul with volitional power even to magnetic rock, because it could move pieces of iron. This position later became known as *hylozoism* (from the Greek words 'matter' and 'alive'). Thales "regarded God as the intellect (or mind) of the universe and thought the whole to be animate [...] and full of deities"; see Schrödinger (1948, p. 65).

Philosophers **Baruch Spinoza** (in a work published just after his death in 1677), and **Arthur Schopenhauer** (published in 1859) developed standpoints that are at least somewhat related to the intrinsic nature position. Summaries of their distinctive proposals may be found in Cottingham (2008, pp. 227-9 & pp. 236-9 respectively).

Recent history

George Romanes wrote a paper in 1885 setting out the intrinsic nature position exactly, though using different terminology. A posthumous publication containing his ideas is freely available from the Internet (Romanes, 1895).

Bertrand Russell, in his *Analysis of Matter*, after stating that from science one can infer nothing as to the intrinsic nature of matter, wrote the following:

> There is no theoretical reason why a light-wave should not consist of groups of occurrences, each containing a member more or less analogous to a minute part of a visual percept. [... Only the abstract mathematical properties of light are scientifically knowable ...] To assert that matter *must* be very different from percepts is to assume that we know a great deal more than we do in fact know of the intrinsic character of physical events.
> (Italics original, my precis in brackets, Russell, 1927, p. 263)

He goes on to say:

> The gulf between percepts and physics is not a gulf as regards intrinsic quality, for we know nothing of the intrinsic quality of the physical world, and therefore do not know whether it is, or is not, very different from that of percepts.
> (p. 264)

He concludes by stating that a tentative argument can be made to the effect that, because "percepts are part of the physical world," this suggests that percepts and the intrinsic nature of the physical might have similarities; or at least "not a complete unlikeness" (p. 264). This typically English turn of phrase, with its double-negative, is a much weaker claim than the intrinsic nature position.

In the end, Russell opted for *neutral monism*, the claim that the ultimate 'stuff' of reality has both mental and physical properties; but it is of itself neutral between mind and matter. This position has a couple of difficulties. First, it seems impossible, almost by definition, to form any concept of this neutral stuff. Second, the physical properties seem to render the mental properties superfluous (or perhaps vice versa). In my opinion these difficulties make neutral monism less attractive than the intrinsic nature position. Obscurity and superfluity are worse faults than that of seeming (at first sight) to be incredible.

Arthur Stanley Eddington sets out his intrinsic nature position in this way:

> In science we study the linkage of pointer readings with pointer readings. [...] *There is nothing to prevent the assemblage of atoms constituting a brain from being of itself a thinking object in virtue of that nature which physics leaves undetermined and undeterminable.* If we must embed our indicator readings in some kind of background, at least let us accept the only hint we have received as to the significance of the background – namely that it has a nature capable of manifesting itself as mental activity.
> (Italics original, Eddington, 1928, p. 260)

He goes even further, attributing mind to all matter, and noting (as Russell did above) that this is a philosophical rather than a scientific question:

> I will try to be as definite as I can as to the glimpse of reality which we seem to have reached. Only I am well aware that in committing myself to details I shall probably blunder.

> [...] The recent tendencies of science do, I believe, take us to an eminence from which we can look down into the deep waters of philosophy; and if I rashly plunge into them, it is not because I have confidence in my powers of swimming, but to try to show the water is really deep.
> To put the conclusion crudely – **the stuff of the world is mind-stuff**.
> (Emphasis added, Eddington, 1928, p. 276)

Galen Strawson has discussed Eddington's argument extensively in a paper, "Realistic Monism: Why Physicalism Entails Panpsychism" (2006). According to Strawson, physics has a double aspect: all that is extrinsically energy is intrinsically mind. His position will be presented in detail in the next chapter.

David Chalmers' philosophical method is to survey the attractions and difficulties of the many positions that can be taken regarding the mind-body problem. For this reason, he does not endorse any specific theory, but he views the intrinsic nature argument favourably:

> There is only one class of intrinsic, nonrelational property with which we have any direct familiarity, and this is the class of phenomenal properties. It is natural to speculate that there may be some relation or even overlap between the uncharacterised intrinsic properties of physical entities, and the familiar intrinsic properties of experience. Perhaps, as Russell suggested, at least some of the intrinsic properties of the physical are themselves a variety of phenomenal property? The idea sounds wild at first, but on reflection it becomes less so. After all, we really have no idea about the intrinsic properties of the physical. Their nature is up for grabs, and phenomenal properties seem as likely a candidate as any other.
> (Chalmers, 1996, pp. 153-4)

Concrete reality characterised

We have intuitions about what it means for something to concretely exist. I am going to say that a putative entity *intuitively exists* if it meets our, admittedly inadequate, pre-theoretic notions. These notions are somewhat inconsistent, as may be seen for example in the section **Empirical reality versus concrete reality** above.

In this section I am first going to give a semi-formal characterisation of what it is for an entity to be **concretely real**. My next task will be to show that concrete reality matches well with intuitive existence in most respects. Moreover, where there are discrepancies, I will provide motivations as to why these are unimportant; and as to why, in such instances, we should ignore intuitions, and use the formal characterisation. (One motivation can be given immediately: discrepancies are inevitable because intuitive existence is not a clear, self-consistent concept.)

From the discussion of the entire chapter, and in particular the previous section, it is evident that an initial plausible candidate for characterising concrete reality should be a refinement of the generic intrinsic nature position. My characterisation is therefore as follows:

A putative entity is *concretely real* if and only if

- it is a *centre of experience* that can *perceive* other concretely real entities
- it has the ability to *act*, which means: (1) to change its current percept to a limited extent; and (2) to change the percepts of some other concretely real entities

Call concretely real entities *experients*, for short. This characterisation contains nothing that is not lucid and familiar. We know what it is to be a centre of experience, because that is what we are; and we know what it is to perceive other entities. When we act to raise our hand, each of our neighbours witnesses this as a change in their percept. Human babies learn to distinguish between their own bodies and their environment by discovering which elements within their percepts are under their control. This characterisation of concrete reality is entirely mentalistic in character.

Thus, an entity is concretely real if it meets Descartes' criterion for existence: "I think therefore I am." This characterisation thus cannot immediately be seen to be absurd; moreover, I contend that it is rational and justified. It is not necessary for the entity to possess any cognitive powers: most entities will be extremely primitive centres of experience.

A *universe* consists of nothing other than such interacting, concretely real entities. There is the further condition that the universe cannot be divided into two or more portions consisting of centres of experience that never interact. If this condition is not fulfilled, then we have several universes.

Any universe composed entirely of experients will also be said to be *concretely real*. It will be called this even if it is not itself – considered as a totality – a centre of experience. The question as to whether or not a universe is itself an experient will be considered separately.

One can think of concretely real experients as being an alternative to, and as a fleshing out of, Mill's skeletal PPobP. This characterisation implies *panpsychism*, in the broad sense that mind exists everywhere in our universe. But it goes further than that: If this characterisation of concrete existence is accepted (and no acceptable alternative has ever been offered), then it is not even coherent to conceive of a universe that totally lacks mind. In particular, it is not coherent to think that our universe existed concretely for the first billion years of its history in the absence of any mind.

Berkeley certainly agreed that the idea of existence without mind is incoherent. He wrote:

> It is very obvious, upon the least enquiry into our own thoughts, to know whether it is possible for us to understand, what is meant by the *absolute existence of sensible objects in themselves, or without the mind*. To me it is evident that these words mark out either a direct contradiction, or else nothing at all. And to convince others of this, I know of no readier or fairer way, than to entreat they would calmly attend to their own thoughts: and if by this attention, the emptiness or repugnancy to reason does appear, surely nothing more is requisite for their conviction. It is on this therefore that I insist, to wit, that the absolute existence of unthinking things are words without meaning or which include a contradiction. This is what I repeat and inculcate, and earnestly recommend to the attentive thoughts of the reader.
> (Italics original, Berkeley, 1710, ¶24)

Although my broadly panpsychic metaphysical system is vastly different from Berkeley's, I agree with every word of the above quotation.

Is this characterisation of concrete reality reasonable?

I believe so. Although it is somewhat vague, the terms in which it is expressed (**centre of perception**, that can **perceive** and **act**) are familiar enough for us to work with. There can be little question that any universe, which by hypothesis meets this description, indeed intuitively exists.

But some things we might imagine as actually existing in fact cannot. Consider a twin of our universe, having exactly the same laws, and beginning in exactly the same physical state. As it develops, this universe gradually diverges from ours, due to the randomness of physical events, in such a manner that life never evolves on twin Earth. We do not know whether or not life has evolved anywhere else in our universe but, if there are such locations, we can likewise suppose that chance events prevented life developing in the twin universe. Without self-contradiction, we can make the assumption that, within the twin universe, life never evolves anywhere throughout its entire history.

The standard view is that mind can only exist where life exists. On this view, the twin universe is entirely devoid of mind. Can we **really** conceive of such a mindless twin universe as actually existing? If we make the effort, I think that the best we can imagine is an abstract mathematical model. Even if we sprinkle Mill's PPobP appropriately throughout the model, this is useless for our understanding. In the mindless twin universe, PPobP is an utterly obscure faculty that is never used.

Bertrand Russell wrote that "we do not suggest there is any impossibility about unperceived existents, but only that no strong ground exists for believing in them" (1927, p. 213). In contrast, my guiding principle is that **the onus is on those who believe in unperceived existents to give a clear account of them**.

It is well known that, whenever we try to formalise our intuitive ideas, we are forced to amend them. As long as the changes are minor, or lead to fruitful progress in our understanding, it is agreed they are desirable. For example, atoms were once thought of as being indivisible. Now we know of subatomic particles, and that atoms can be divided in specific situations. For many purposes, atoms can still be regarded as being indivisible.

Something very close to the twin universe can concretely exist according to my characterisation of the latter term. Roughly speaking, we amend the twin universe by sprinkling experients, in precisely those locations where PPobP was sprinkled previously. Now mind – albeit extremely primitive – exists everywhere, and it never develops cognitive powers. As previously argued, there is no absurdity in asserting that, thus amended, this twin universe concretely exists. Even more beneficial, **we now have a clear, self-consistent, definite, and**

adequate *concept*, as to what the concrete existence of entities within this twin universe – and indeed *any* universe – amounts to.

Is this characterisation necessary?

While not proof can be given that the characterisation of concrete reality presented here is the only possible formalisation of intuitive reality, there are several points in its favour:

- Substance monism

This desirable feature asserts that there is only one kind of 'stuff' in the universe. We know that we (and our pets) are experients. If most matter is wholly insentient, this takes us close to the awkward position of Cartesian substance dualism.

- The evolution of mind

It is clearly much easier to explain how complex experients possessing cognitive powers evolved from simple experients without cognitive powers, rather than from insentient matter. Admittedly, a detailed explanation is still called for, but human consciousness will no longer be a "Hard Problem". According to David Chalmers, the *Hard Problem* is that of explaining consciousness in a manner which accepts the fact of consciousness – **as experienced by us**; and which also respects current science (1996, Introduction).

Chapter 10 will show how experients can combine over time to form more complex experients that are able to engage in more complex behaviour. This, at least in principle, gives the first step in explaining the evolution of consciousness.

- Concrete reality is a more precise specification of intrinsic nature

As already seen, we have excellent grounds for believing that some form of the intrinsic nature position is true.

Moving on

The following chapters will develop these ideas, including showing: how physics can be characterised in entirely mentalistic terms; the essential role mind has in the causation of the universe; and much else. In the course of this, the specifics of my characterisation of concrete reality will be discussed and justified.

Chapter 8

Pan-idealism

The great majority of contemporary philosophers studying the mind-body problem hold to *physicalism*, in which the early universe was totally insentient, and mind only arises within a few complex physical (specifically biological) systems. One motivation is that many – wrongly as I hope to demonstrate – believe that physicalism is the only metaphysical position consistent with science. Almost all scientists agree; Stephen Hawking for instance explicitly held this view. But, as discussed in chapter 2, the history of attempts by physicalists to give any minimally acceptable solution to the mind-body problem has been that of consistent failure. The implausibility of giving any complete and rational reduction of mind into physical terms (say by claiming to identify pain with a certain pattern of neural firings); likewise, the vacuity of 'emergence' as a putative explanation of the arrival of mind in complex physical systems: these and other failures count strongly against physicalism.

Because of these difficulties about five per cent of philosophers have adopted the position of *panpsychism*, holding that every physical thing has mental properties in addition to physical properties: mentality (for the most part extremely primitive) exists throughout the universe. One powerful motivation for this position is the intrinsic nature argument, which has just been made at length. Another factor is the lack of any rational account as to how mind could credibly emerge from insentient matter. There are other arguments.

Pan-idealism is a more radical metaphysical position than panpsychism. It holds that everything that concretely exists ***is* a mind**, pure and simple. The cosmos is nothing other than an interacting collection of minds. This position is thus a variant of idealism, but one that is fully realistic about the contents of the universe. This chapter explores pan-idealism, whilst postponing discussion of how it fits in with the specific physics of our world.

Pan-idealism characterised

Pan-idealism is very much a minority position, being held by only a handful of people (under various names). A few advocates are Romanes (1895); James (late in his life; 1909); Bolender (2001); and Mexner (see Brüntrup & Jaskolla, 2017, chapter 16). My own specific contribution is to tie pan-idealism as closely as I am able to objective physics. This is done throughout Part III, but especially in chapter 10.

Here are the postulates of pan-idealism:

Assumptions about the physics of our universe

Pa. Our universe has an objective physics

Pb. This objective physics includes a catalogue **C** of all of the *true individuals* in the universe (at least as it exists up until the present cosmic time)

The catalogue would contain all individual instances of each type of entity. For example, this particular molecule and that particular cat are true individuals, and so are in **C**. Catalogue **C** contains, among other things, all elementary particles, fields, atoms, molecules, cells, organs, organisms.

Pc. Aggregates of true individuals

Considered as a single thing, a miscellaneous collection of objects – such as a **place-setting** consisting of a knife, fork and spoon – is not a true individual. Instead, it is merely an *aggregate* of true individuals. All aggregates can be broken down into components that will consist (at some lower level – certainly at the molecular level) of true individuals.

Pd. All of these things (both individuals and aggregates) are concretely real

Assumptions about the omnipresence of mind

Mind is present throughout the universe in the following sense:

Pe. Every true individual in the universe is a centre of experience (an *experient*)
Pf. Every experient has a qualitative *percept* of at least some other experients
Pg. They behave lawfully, with a degree of freedom, according to their percepts

To give a simplistic example: A *free particle* (i.e., one that is not entangled with anything else) is a centre of experience. If it is a participant in a Young's Slits experiment, it has an intrinsically qualitative percept of the entirety of the apparatus, and it has a certain limited, *lawful agency* enabling it to opt where it will land on the target screen.

A particle's propensities to make certain choices are reflected in its percept of the specifics of the entirety of the experimental setup – whatever this happens to be when the particle passes through. See Figures 4.1 and 4.2. Figure 4.2(a) gives the particle's propensities when either the left slit or the right slit is open. Figure 4.2(b) gives the propensities when both slits are open. (A more rigorous account of lawful agency will be given in chapter 10.)

In asserting **Pg**, I am certainly not claiming that particles, molecules, and other primitive entities can think: they have no cognitive powers. Nor does it amount to the claim that the kind of basic freedom that these entities possess (which is no more than the capacity to do this or alternatively to do that) amounts to free will. This will be clarified in the next chapter.

> **Ph.** Experients can exist in *hierarchies*: an *organic individual* is one that sits above others in a hierarchy

Organic individuals include atoms, molecules, cells, organs, and animals. But physical ultimates (such as electrons) are not organic individuals; and neither are aggregates.

The assumption of pan-idealism

The above postulates make it clear that mind is omnipresent. The following assumption distinguishes pan-idealism from panpsychism (in which some physical facts are fundamental):

> **Pi.** *Pan-idealism*: All of the truths of objective physics can be fully reduced to facts about experients, their percepts and their volitions

Strictly speaking, assumption **Pi** is a promissory note that the pan-idealist is obliged to try to fulfil; rather than a postulate that is to be accepted without justification.

According to **Pi**, we no longer expect even the truths of objective physics (which are by definition fundamental **within the domain of physics**) to be **absolutely** fundamental. For example, suppose Einstein's well-established $E = mc^2$ remains true in objective physics; then the pan-idealist is under an obligation to show how this law can be expressed in terms of facts about experients, their percepts and their volitions. In order to solve the (pan-idealist variant of the) mind-body problem, the pan-idealist must attempt to make this reduction for the entirety of physics.

Taking pan-idealism seriously

There are two general categories of philosophers. Nowadays, the majority are content to analyse and refine metaphysical systems. They do this without committing themselves to one particular system. Indeed, these analytic philosophers see this as a virtue.

In contrast, traditional metaphysical philosophers commit themselves to a particular metaphysical system that they believe is most likely to be true of the world. They attempt to present this system in as accurate and detailed fashion as they are able, and to refine it. Of course, as with scientists, it is essential to have the flexibility to reject a theory if it becomes untenable. Nowadays moreover, a crucial necessity for all philosophers is to ensure consistency with present-day scientific understanding, especially with physics.

Pan-idealism is a metaphysical position that, to the best of my understanding, I take to be literally true; it is not just a fanciful intellectual notion. Moreover, I take pan-idealism to be true, not just for our universe, but for any other concretely-existing cosmos (if there be any), no matter what physics it has. My reasons for believing this are: first, as will become clear, pan-idealism gives a lucid and rational account of the notion of concrete reality; and second the absence of any alternative account of this essential concept, as was discussed in the previous chapter.

This does not mean that pan-idealism is complete as a theory – far from it. Nor is this an exhortation for you to agree with me – you will make up your own mind. Having said this, I would point out that, in order to assess pan-idealism impartially, it is necessary to properly evaluate certain matters:

First, it is crucial to put aside physicalist presuppositions. This is because it is very easy to reject pan-idealism on the false basis that it contradicts some physicalist assumption, when this assumption is plainly false under pan-idealism. (For example, the assumption that physical effects have physical causes. Another such assumption is that mental events (say tastes) must depend on physical events (say neural firings).)

Second is the charge of absurdity: Isn't it absurd to assume that the ultimate entities of which the universe is composed are in truth centres of experience? The intrinsic nature argument (chapter 7) shows that is a rational position. Isn't it absurd to assume that these entities possess volitions? The truth is that, if consciousness has no effect whatsoever, then this calls the very existence of consciousness **as such** into question; and this is an even greater absurdity. Moreover, on the other arm of the weighing scale, the longstanding, utter failure of physicalism to give a rational account of mind must also be considered a hefty absurdity.

Third, doesn't pan-idealism contradict science? No: Assumption **Pi** – which the pan-idealist is under an obligation to justify – requires pan-idealism to be consistent with objective physics, and hence with present-day science – at least insofar as this is correct. The latter qualification merely reflects the familiar truth that all scientific statements are in principle falsifiable; it is not meant to deny that we should have the greatest confidence in science.

Variants of pan-idealism

The postulates of pan-idealism leave several questions open. These are best solved by two methods: engagement with contemporary science; and conformity with desirable philosophical principles. Possible variants of pan-idealism will be touched on here, in the order in which the postulates **Pa** through **Pi** were given:

The contents of the universe (Pb)

For the most part, it will be scientists who decide what belongs in catalogue **C**, based on the best current physics. But it will be open to pan-idealists to have some input into the discussion because they make the additional metaphysical claim **Pe** that every true individual is a centre of experience. For example, is spacetime, considered as a whole, one unified centre of experience?

True individuals versus aggregates (Pb & Pc)

How does one distinguish between true individuals and an aggregate of individuals? This is a problem for panpsychists and pan-idealists alike. Clearly the above-mentioned place setting is an aggregate; but what of a rock?

Often, such answers are given out of squeamishness – a fear of being accused of advocating foolishness. But it is better to take a position grounded on dependable philosophical principles, however outlandish it may appear at first sight. In the specific theory I present in detail later, rocks are indeed centres of experience, albeit having zero cognitive power.

Experients can exist in hierarchies (Ph)

How is this possible? This is a well-known and thorny problem for panpsychism, known as the *combination problem*. It was first pointed out by William James in his seminal work that founded the science of psychology (1890). In a section entitled "Self-compounding of mental facts is inadmissible", James pointed out that if one adopted panpsychism, which he called the theory of "mind dust", it seemed impossible to explain how such particles could amalgamate in order to become a unitary human mind (1890, pp. 160-4). No fewer than five chapters of Brüntrup & Jaskolla (2017, their Part III) are devoted to this problem.

In chapter 10, I solve this problem. It will turn out that each hierarchical system of experients 'corresponds to' an entangled system of particles. More accurately, a hierarchical system of experients **is perceived by other experients as being** an entangled system of particles. This solution to the combination problem is a key reason for giving this book serious consideration.

Are experients long-lived or transient?

At first sight, it is possible to conceive of one's consciousness in two general ways: You might think of your mind – at least while awake over the course of the day – as a unitary thing, a continuous thread of experience that endures for roughly 16 hours until sleep intervenes. This thread is akin to a short portion of your world line.

Alternatively, you might think of your experience as a succession of moments of consciousness – each lasting a fraction of a second: What am I perceiving **right now**? Rather than being a single continuous thread, according to this latter sketch each day is more like the fine chain of a necklace, each link of which lasts but a moment, and each of these links intertwined with the next.

It would be possible to have variants of panpsychism loosely based on either of these preliminary models. A single experient could last several hours, or a fraction of a second. This chapter will explore the second idea, as developed by William James.

These are only primitive initial ideas. In a more accurate model, a human person comprises a multitude of interacting threads/chains of experients, of greater or lesser complexity; all of which interact with one another.

Pan-idealism's unique mind-body problem

It cannot be stressed too strongly that the mind-body problem for pan-idealism is **unique**. For all variants of physicalism (including panpsychism), a putative solution to the mind-body problem is an attempt to give an explanation of all mental facts in physical terms. In contrast, a solution to **the mind-body problem for pan-idealists** must explain how to give a complete account of the objective physics of the cosmos in entirely mentalistic terms. This explanation must be lucid, rational, and consistent with science. Thus, the mind-body problem is turned upside-down: Mental facts are taken as fundamental and given. Physical facts are derived entirely from these.

It might be objected that the first two postulates of pan-idealism, **Pa** and **Pb**, are physical rather than mental in character. The brief answer is that the **only** way we can gain any knowledge about the universe – whether scientific or commonplace – is through our experiences. We can leverage this indisputable truth to **identify** all physical facts with facts about experiences (not only our own experiences, but those of all other experients). The rest of the chapter outlines step by step how this is done.

Length in pan-idealism

The task is to show that physical properties are nothing over-and-above mental properties; and we start with length. Suppose two scientists are measuring the length of a bone. The woman places a 10 cm ruler alongside the bone. She perceives that the ends of the ruler each happen to lie adjacent to an end of the bone. (For simplicity – in a manner analogous to Einstein's measuring rods – we're assuming that the bone and ruler happen to be equal in length.) Based on her percept, she concludes that the bone is 10 cm long.

She asks her colleague to verify this conclusion. He also perceives that each end of the bone lies beside each end of the ruler, and so he agrees that the bone is 10 cm long. It is well known that all scientific facts have to be – at least in principle – verifiable. What is less often stressed is that verification is **of necessity** carried out by confirming that human subjects have compatible percepts – no alternative method of verification is conceivable. Let us call this verification condition *intersubjective consilience.*

A candidate definition is that *length* is identically the maximal intersubjective consilience between the percepts of human observers. As it stands, this suggestion would be hopelessly anthropocentric: the reality of the property of length should not depend upon the existence of humans. However, we can correct this, in a manner befitting pan-idealism, by extending the condition to **all** experients. For example, a dog, although its ability to think is limited, has a mind and perceptual apparatus that are fairly similar to those of humans: the images of the bone and the ruler take up equal portions of the dog's visual percept. Its experiences as it wanders around these items are empirically consilient with the finding of the scientists that the ruler and bone are equal in length. The dog's unthinking percept is an effective measurement of the length of the bone – even if the scientists were to leave the room. Although we do not have epistemic access to the dog's percepts, we may be very confident that they do occur as described – they are part of the ontology of our pan-idealist universe.

We can extend this to even humbler experients. For example, pan-idealists maintain that even molecules (say in the air) have percepts containing information about, among other things, the sensed locations of the ruler and bone. These percepts – which we have good reason to believe in by the intrinsic nature argument – are likely to be extremely primitive. Although they are not directly knowable, they could perhaps one day be deduced theoretically.

Definition: the *length* of an object is obtained by maximally combining (and reconciling as outlined above) structural information from the percepts of **all** of the experients that perceive it – including the experients that make up the object.

This definition is intended to be an **identity**: length is **nothing other than** the best possible reconciliation of structural information taken from the percepts of numerous experients. This maximal intersubjective consilience is not intended to be a practicable method that will enable a scientist to determine a particular length, because no human can have information about the percepts of all experients. It is sufficient for our purpose that, as a metaphysical and mathematical fact, such a length exists.

Is the above definition consistent with scientists' usual concept of length? Science is entirely empirical in its methodology – even its theoretical findings are ultimately based on human perceptions (observations and measurements) and volitions (to carry out experiments). If a bone has a certain length as defined by pan-idealism, then it should have the same length according to scientists (all of whose streams of consciousness are sequences of experients). We can assume that usually the percepts of scientists are a representative selection from the percepts of **all** experients. Why representative? Human beings would be unfit in evolutionary terms if their percepts did not (for the most part) accurately reflect their situation.

Identity

Length is thus identical to reconciled information derived entirely from the domain of mind – it is consistent with the idealistic postulate **Pi** of pan-idealism. There is no need to suppose that there is any such thing as an additional and obscure 'real physical length' that somehow exists **'out there'**. It is tempting for me to wave my hands at the ruler and bone whilst I utter the latter, emphasised words; but in doing so I merely perceive my hand waving within the totality of my visual percept. Likewise, each of my companions has a percept of my hand waving within the totality of his or her visual percept. There is no such place as 'out there' that is 'external' to any percept. Having said that, my hand indeed concretely exists, consisting as it does of innumerable experients.

The Moon and space

To give another example, the Moon and the particles that constitute it are always experiencing and being experienced by one another. The Moon therefore existed in a clearly defined manner – with a precise diameter – even in the early universe before the advent of life.

Once length has been reduced in this manner to mind, it is easy to convince oneself that the concept of space has, by that fact, been reduced. (According to current physics, even the vacuum of outer space teems everywhere with short-lived entities; which pan-idealism identifies as experients.)

Time in pan-idealism

Pan-idealists must also be able to characterise time – as this is known to physicists – in terms of the subjective time possessed by experients of greater or lesser complexity. This reduction can only be outlined tentatively and incompletely, but the same principle of maximal intersubjective consilience will again be used. The procedure given here for identifying time in terms of streams of consciousness is essentially that followed by Foster (1982, pp. 257-61) in the different but related context of defending Berkeley's idealism.

We may suppose that experients are short-lived, at least in the human case. How are these 'moments of experience' to be connected together into a 'stream of consciousness'? William James provides the answer: a moment of experience is not a temporal instant but rather covers a brief period during which events A B C D are perceived to occur successively within this one, indivisible experience. Within each human experient, in other words, there is already a perceived passage of time. James calls the duration of the present moment of experience the *specious present* (James, 1890, p. 573). So, a human stream of consciousness has the following schematic form:

In experient #1: A B C D are perceived successively
In experient #2: B C D E are perceived successively
In experient #3: C D E F are perceived successively

In experient #4: D E F G are perceived successively

C is the same event which occurs in the percepts of the first three experients; similarly for the other events, see James (1890, p. 571; & 1909, pp. 67-68). Taken together, experients 1 through 4 therefore make up a single stream of experience in a pan-idealist universe. This stream of experience may loosely (omitting mention of the experients) be referred to as 'A B C D E F G'. I will call this process of connecting together moments of experience into a stream of consciousness *daisy-chaining* (even in the non-human case).

It might be objected that scientific practice characterises time solely as a sequence of objective events A, B, C, D, such as the ticks of a clock, without any reference to subjective experience. This not the case, however, because, in order for a human even to comprehend what a clock amounts to – as an object in which 'something is going on' – he or she must already have some minimal subjective sense of the passage of time. If humans did not have this subjective temporal sense, then all of the ticks would be 'heaped together' uselessly when they occur within a single percept (see James, 1890, p. 571).

Higher animals presumably have similar streams: it is a dog's subjective stream of experience (and memories of previous chases) that helps it to catch a rabbit. As one goes down the phylogenetic scale, it becomes less clear that the above model of a stream of experience applies. The specifics of applying this model to an ultimate such as an electron are completely unknown, but the form of pan-idealism advocated here holds that even an electron is in fact such a stream of experience, made up of daisy-chained moments of experience, in the course of which perceptual events occur as above.

How are distinct streams of experience to be combined into a single physical time? Again, we use the principle of maximal intersubjective consilience. Suppose two streams of experience contain the following sequences of events (the experients themselves have been omitted):

Stream one: A B C **D E F G** H I J
Stream two: P Q R **d e f g** S T U

We suppose that events A B C are nothing like P Q R, and events H I J are nothing like S T U, but that the corresponding events in bold are similar to one another: event **D** is similar to event **d**; event **E** is similar to **e**; likewise, pairs of events **F f** and **G g** are similar.

The natural interpretation here would be that two streams of experience existed well apart until events **D** and **d**; they stayed close together until events **G** and **g**; and then they went their separate ways. To give the simplest possible example, suppose two people come together briefly to look at a digital clock. Events **D** and **d** are the clock's display changing from 1 to 2, events **E** and **e** are the clock's display changing from 2 to 3, and so on. **D** and **d** are not quite identical because the clock is seen from two different viewpoints. The two streams of experience then move apart. In such situations, I will say that the two streams of experience are *linked* between the specified events (here between events **D** to **G** and **d** to **g** respectively). While they are linked, they share a common perceived flow of time.

To sum up so far: (1) The universe is everywhere comprised of a multitude of experients. (2) These are daisy-chained together, by means of the common perceptual events that occur within them, into streams of experience. (3) Streams of experience are then compared, and are occasionally found to be linked to one another, temporarily sharing a perceived flow of time. The final step is (4): If the structure of each experient's percept is sufficiently rich (containing many events, even in the case of a physical ultimate) then the linking that maximises consilience should be well-defined across all experients throughout the universe. In pan-idealism we take this maximal consilience to **define** *physical time*. In other words, physical time is nothing other than the fact of this maximal intersubjective consilience.

The construction made here is quite in keeping with the notion of time described in special and general relativity. This is because two streams of consciousness, belonging to the maximally consilient linkage (as in Streams one and two above), can be regarded as co-moving during the period when their respective percepts are similar. Moreover – for any theory of physics – the definition of time given here is fully consistent with what scientists must actually do: in reality they cannot avoid ascertaining time empirically; that is to say, in terms of their common experiences. The only difference here is that scientists are now regarded as a representative subset of all experients.

We will never be able to carry out the programme described here – even in principle – because we are not omniscient. Nonetheless, there can be no logical objection to the existence of a universe consisting solely of experients, whose percepts are **in fact** linked together in such a well-defined manner; so that it instantiates – as derived, secondary properties – both length and time.

Other physical properties in pan-idealism

Other physical properties depend upon the specifics of the objective physics of our world. According to the *Bureau International des Poids et Mesures* (BIPM), the international agency that regulates SI units, there are currently seven base units from which all other units are derived (BIPM, 2019, 9th edition). The base units are themselves in turn derived from certain observed physical constants, called *SI defining constants*. These include: c, the speed of light in a vacuum; h, the Planck constant; and e, the elementary charge (the charge of an electron is minus one).

SI units are somewhat human-centred: the speed of light is

299,792,458 metres per second for instance, and both the metre and the second were chosen for human convenience. We can arrive at more natural units by choosing them in such a way that all of the defining physical constants are equal to 1. In the case of general relativity this is often done implicitly for *c*, where the slope of the light cones is depicted as being 45°: see Figures 3.2-3.5.

Particles of various types are distinguishable in our pan-idealist universe. They differ in their different spatiotemporal properties and behaviours – as these manifest themselves in our percepts. According to pan-idealism, these structural differences are also present in the rudimentary percepts of even primitive experients. In the very early universe, for example: (1) the world lines of photons formed themselves into light cones; also (2) the world lines of electrons always remained within these light cones. These last two facts (and all physical facts) boil down to mathematical truths about consilience between the percepts of those experients which existed in that era.

To sum up: **Everyone agrees that** all physical properties can be expressed in terms of observed spatiotemporal properties and goings-on. **Pan-idealists alone assert that** all physical properties are nothing other than mathematical facts about the combined percepts of experients (which are omnipresent).

Reducing physical causation to mental causation

The next task is to show that physical causation is nothing over-and-above mental causation. The model for the dynamic of the universe is given by human agency. I am in a certain mental state in which I perceive that my hands are by my sides. I have a volition (I decide) to wave to a friend, and I then perceive that I am waving. My friend, and anyone else in the vicinity, can see me perform this action. This dynamic is entirely mentalistic in character.

In the same way, ultimates, such as electrons, have a certain degree of agency, with lawful freedom to act according to their percepts. Any two electrons having identical percepts will have identical predispositions to act in each possible manner – if the probability to land at A rather than B is ¾ for one of them, then it will be ¾ for the other. This **mind-like in character lawful agency** is the ultimate grounding of all causation in our universe, and it underlies the laws of physics (and also so-called 'physical causation'), which are nothing other than the perceived structural/mathematical regularities of nature.

The laws of physics have straightforward mathematical formulae. This is because: all physical ultimates are extremely rudimentary in terms of their percepts and volitions; moreover, all physical ultimates of one particular type (say all electrons) are identical in terms of (1) what they perceive in any given environment, and (2) their propensities to act.

Pan-idealism, at least as discussed in this book, has the consequence that two humans who are absolutely identical as hierarchies of experients (this implies they have identical bodily, brain, and mental states), and who are in identical environments, must have identical propensities to choose A rather than B. (This situation is utterly impossible in actuality, I know.) It follows that, at least in principle, the universe is lawful at the biological level. Human agency will be discussed much more fully in the next chapter.

Causation in pan-idealist universes

There are certain rules that apply to all 'reasonable' pan-idealist universes.

- No epiphenomenalism

This is a robust metaphysical principle that is invaluable for any metaphysical system. In pan-idealism it prevents us from exaggerating the perceptual powers of experients. For example, animism is not permitted. The rule affirms that every feature within the percept of any experient must, under at least some circumstances, play a role in its behaviour.

Even though we do not have direct access to the percepts of any other experient, we receive indirect clues about them because they affect its behaviour. This is what lies behind our empathetic access to the minds of other experients, whether human, other animals, or even (I would claim) the limited mentality of rudimentary quantum systems.

- Pan-idealist causal closure

This states that there can be no change in the percept of any experient that is not caused by the agency of some other experient. In other words, there can be no causation that does not originate from within the universe.

This is analogous to the causal closure rule pertinent to a physicalist universe, and it is equally compelling. The only differences between these rules are due to the different ontological foundations (experiential versus physical) of these two systems.

- Causally connected

There are direct or indirect causal (agentive) connections between all experients. If this were false, then we would be considering several distinct and independent universes.

From now on we will assume that every pan-idealist universe possesses these essential characteristics.

Objective physics

To sum up: the ontological base (what fundamentally exists) in pan-idealism are the experients, their percepts, and their agency. If there were no communalities between the percepts of the experients, then there would be no reason to believe that they exist within one universe; instead, they would merely be a hodgepodge of unrelated experients. In any universe, we can assume that communalities exist.

Objective length and time are physical facts. Each is defined, as described earlier, as being identical to a specific type of ***maximal intersubjective consilience*** between the percepts of experients. All other physical facts – indeed the entirety of objective physics – are characterised in the same way: by means of maximal intersubjective consilience.

The objective physics of a pan-idealist universe is unequivocally true; nevertheless, all physical facts are **secondary**, because they are, in principle, completely reducible to experiential facts.

There is no need for the objective physics to be known, or even knowable. Maximal intersubjective consilience is **not a process** whereby one seeks to find the maximum (even though scientists, of necessity, often develop their theories in this manner). Instead, **objective physics is a logical truth** – a mathematical consequence of the ontological base of the specific pan-idealist universe under consideration. A simple example of the objective physics of a toy universe will be given in chapter 11.

The advantages of pan-idealism

Pan-idealism gives clear answers to many formerly intractable problems in the philosophy of mind:

- *The explanatory gap*: How can mind and matter be identical when they have such different properties?

In pan-idealism, everything physical (entities, laws, properties and causation) can be reductively identified as mental dittos, by using the principle of maximal intersubjective consilience.

There can be no explanatory gap in pan-idealism because **we do not know anything about the physical except through our experiences**.

- *Mental causation*: How can mental goings-on have physical effects?

Jaegwon Kim has argued forcefully and in detail that, under the assumption of physicalism (1996, pp. 9-13), the problem of mental causation is a "profound dilemma" (p. 236). He holds that a physicalist must accept the causal closure of the physical world (no physical events have a non-physical cause) on pain of becoming a Cartesian dualist (pp. 232-3). The problem for the physicalist is how to fit mental causation ('I took an extra bite of chocolate cake because it tasted so good') into the physicalist picture. (In later editions of this book, Kim moves to a more mainstream, but in my view less interesting, position. He asserts that everything about consciousness can be explained except for qualia. However, disregarding qualia is, as Strawson rightly argues (2009a, p. 35, note 6), the denial of consciousness.)

In pan-idealism, as has been discussed above – far from assuming the causal closure of the physical world – there is no physical causation at all; and the only dynamic is volition based upon perception. In pan-idealism, the fundamental ground of all existence is wholly mentalistic, and all physical realities are entirely derived from this ground. What seems at first sight to be physical causation consists solely in the volitions of experients as reflected in the percepts of other experients.

- *Perception*: How do I have a mental percept of the physical world?

This is exemplified by David Hodgson, "Light reflected from my red pen is focussed by the lenses of my eyes on to the retinas:

and this results in electrical-chemical signals going to my brain, and then in further electrical-chemical processes within my brain: and *I see the red pen*" (emphasis original, 1991, p. 1).

In pan-idealism this process does not involve a mysterious change of ontological category – from physical to mental – at the final step. Instead, every segment in this sequence of events is entirely mental: an act of perception and volition. In other respects, the pan-idealists' account of perception is identical to that of physicalists – in terms of information flow.

- *The problem of qualia*: Recall that *qualia* are things such as 'hot', 'red', 'pain', **as these are experienced by us**. How is it possible that we have such experiences?

In pan-idealism there is no problem of qualia because they are fundamental. Moreover, it is the presence of qualia that distinguishes concrete reality from an abstract mathematical model.

(The converse problem – of accounting for the physical in terms of intrinsically qualitative percepts – has been the topic of this chapter. Structures are abstracted from percepts analogously to the way a vaguely oblong region may sometimes be seen in the qualitative field that is a Mark Rothko painting. Abstraction (literally 'taking away') is not an obscure notion.)

- *The problem of zombies*: Philosophical zombies are putative creatures, physically and behaviourally identical to us, but entirely lacking mind (Chalmers, 1996, p. 94). Two questions are, "Are zombies conceivable? Why are we not zombies?"

In pan-idealism, to exist concretely as an individual is exactly to be a system of experients: my brain (physics and all) is nothing over-and-above a hierarchy of experients. So, take away my mind (in particular, that part associated with my stream of experience), and you have, in that very action, annihilated a chunk of my brain. Zombies are not conceivable.

- *Realism about objects in the world*: Are the objects we perceive in the world real? What do we mean when we claim that an object is real?

In contrast to some variants of idealism, in pan-idealism the reality of physical things is accepted by hypothesis. For the most part, we leave it to scientists to determine what is real, but it is for philosophers to establish the character of these objects.

Moreover, pan-idealists have a far richer notion of the reality of objects in the world than do physicalists. There is general agreement (to the best of our current knowledge) about what physical entities, properties and laws are present in our universe. But, in addition, pan-idealists hold that all individual physical entities are experients; and moreover, all physical properties and laws are determined by the experiences and volitions of these experients.

It is standard (i.e., non-panpsychist) physicalism that has a fatal weakness concerning realism. According to this position, a universe can concretely exist (be instantiated) even in the absence of any experience. But what is it to be instantiated in the absence of experience? Most physicalists say nothing about this problem. Jeffrey Poland states:

> The physical ontological base is completed by a characterisation of the class of all possible total [spacetime] distributions of physical objects and attributes [...] subject to the constraint that the laws of physics are satisfied. [...] One of these distributions [W] will correspond to the current total distribution of such objects and attributes; the others [X, Y, Z...] will be alternative total distributions that define the

> nature and limits of what is physically possible in this world. (Poland, 1994, p. 132; labels W-Z added)

However, Poland says nothing that would distinguish our instantiated world W from merely putative ('imagined') worlds such as X, Y, and Z.

The "ontic structural realism" advocated by Steven French (2014) illustrates how vanishingly thin the physicalists' view of the world is. He hopes to distinguish an instantiated world by the fact that it possesses 'causation', but upon inspection this move is not intelligible. Scientists infer that A causes B solely on the basis of evidence that A is regularly followed by B. The question arises: If we find regular occurrences of what might loosely be described as the structural pattern [A(t), B(t+1)] in a particular mathematical model, then what characteristic (unrelated to experience) distinguishes the universe that instantiates this model from the model itself? (Causation as it is knowable to science cannot help because it is already present in the mathematical model. French's 'causation' must be something else, but what?)

As previously discussed, Max Tegmark recognises this difficulty and asserts that every consistent mathematical structure is an instantiated physical world – a position he calls "Radical Platonism" (2003, p. 14). This seems highly implausible in its profligacy, and also to those of us who are realists about qualitative experiences such as pain.

The burden is on mainstream physicalists to explain how an instantiated but unperceived physical world – which is what they take our very early universe to be – differs from a mathematical abstraction; until they are able do so their position is not comprehensible. In contrast, a pan-idealist holds that an instantiated world cannot be other than one that is everywhere experienced and subject to volition by experients: **experiences and volitions are the very hallmark of concrete existence**.

Some difficulties of pan-idealism

Pan-idealism has some remaining difficulties, but these are few:

- How can we determine intersubjective consilience in practice?

The mathematical regularities between the percepts of **all** experients do exist in principle, as we have seen in the example of length. But a means to calculate intersubjective consilience **perfectly** would only be available to an omniscient observer of the universe – one with complete knowledge of the percepts of all experients. As finite beings, we should not expect to have anything approaching such perfect knowledge. Fortunately (and amazingly), our constricted perceptual grasp of the universe is sufficient for us to attain a superb knowledge of its physics.

- The problem of **specifically** deriving the physical from the mental is as yet unsolved

For example: How does the extremely complex hierarchy of experients constituting the human brain give rise to the specifics of its observed physical structures – its neurons, convoluted surface, and all the rest? Most of this experiential hierarchy is wholly unknown – even our own stream of consciousness forms only a minuscule, and imperfectly known, portion of it. We shouldn't be surprised therefore that the problem is unsolved. The same difficulty applies even to the simplest entities, such as a molecule. In this case, the problem arises because we have no notion of the experiences of such entities, as postulated by **Pe** through **Pg**.

In defence of pan-idealism: the converse problem for physicalists – that of explaining the specifics of mental goings-on in terms of physical goings-on – is equally intractable. The well-known correlations between these types of happenings (such as an experience of cold being associated with a particular pattern of neural firings) are equally relevant for both physicalists and pan-idealists. Such correlations may one day prove to be helpful stepping stones in arriving at the specifics of a solution to the mind-body problem; but they do not, of themselves, amount to a solution.

Moreover, as shown earlier, the mind-body problem is already solved **in principle** in pan-idealism: there is no conceptual difficulty in understanding how the physical arises from the mental. (Contrast this success with the longstanding intractability of this problem for physicalism.)

Strawson's pure panpsychism

The ideas expressed here have antecedents. Idealism was influential in the 19th century and the first quarter of the 20th, when it fell out of fashion. Prominent was the idealism of Schopenhauer (*The World as Will and Idea,* 1844) which took the entire universe to be a unitary entity with a single, blind Will. Our own existences as persons are merely aspects of this Will. In reaction against this, William James, late in his life (in *A Pluralistic Universe,* 1909), while keeping idealism, explicitly rejected Schopenhauer's unitary position. James found it difficult to reconcile his position with the physics of his era, so he boldly concluded that this physics must be in error. These and other idealists were discussed in the last chapter.

For the remainder of this section, I will outline Galen Strawson's pure panpsychism, noting its similarities to and differences from my own position.

Strawson's theory

Galen Strawson (2006b) has developed an argument for *pure panpsychism,* the view that all being is experiential being. His pure panpsychism is, of itself, identical to the pan-idealism that I have described here. However, in order to make this consistent with our scientific knowledge, he felt it necessary to harness pure panpsychism alongside what he calls ESFD monism, a form of dual-aspect monism. I will argue that these two systems contradict one another, and that pure panpsychism alone is in principle sufficient to account for all of the scientific and commonplace facts of the world.

Strawson gives a succinct account of his position in (2009b), and for the most part I will be referring to this. (The same arguments can be found in his more detailed article (2006b). His 2006a paper (reprinted as 2009a) is also relevant.)

Pure panpsychism is the view that all being is experiential being (2009b, p. 57). The universe, in its ontological ground, consists of numerous *sesmets*, which is short for *subject of experience that is a single mental thing* (p. 60), and nothing else. Strawson's 'sesmet' is exactly what I have called an 'experient'. He continues, "[I]t remains central to the present view that the inside of an experience or sesmet…, i.e. its experiential nature, is its whole essential nature, its whole being"; moreover, although sesmets might be said to possess physical energy, this "is not something ontologically extra" (p. 62). Strawson notes that the concept of a subject is implicit within the very notion of experience – essentially because a free-floating taste of chocolate (say) is an incomprehensible notion. Moreover, there is a corresponding 'thin' notion of *subject* which is nothing other than **that which has an experience**. Thus understood, subject and experience are really aspects of what is one thing – a sesmet (pp. 59-60).

Strawson remarks, "As it stands, the thin conception doesn't offer any support to the idea that [sesmets] are short-lived or transient entities. I suspect that they are always short-lived in the human case, as a matter of empirical fact, that the stream of human consciousness is constantly interrupted, in ways large and small, but [long lived or even immortal thin subjects also qualify as sesmets] by the present definition" (p. 59).

The true ultimates of physics, whatever they turn out to

be, whether particles, fields, strings, etc., are in fact sesmets (p. 60). Strawson states, "There is no more difficulty in the idea that ultimate sesmets have sensation and intentionality and represent things than there is in the idea that one particle exerts attribute or repulsive force on another – for these are the same thing" (p. 63). As is standard in panpsychism, Strawson affirms that sesmets can combine, "Some but not all pluralities of sesmets constitute further numerically distinct sesmets" (p. 60). Sesmets can thus sometimes exist in hierarchical systems.

Pure panpsychism is a form of idealism, at least as I have defined the latter term in this book. Strawson, for credible reasons, not least that pure panpsychism affirms the reality of all the things in the world that are normally accepted to exist, prefers the term "mentalism" for his position (p. 61). In my view, this is a linguistic rather than a substantive difference.

ESFD monism

Strawson feels the need to combine pure panpsychism with another position: He summarises *Equal-Status Fundamental-Duality (ESFD) monism* in terms of three premises "[1] Reality is substantially single. [2] All reality is experiential and all reality is non-experiential. [3] Experiential and non-experiential being exist in such a way that neither can be said to be based in or realised by or in any way asymmetrically dependent on the other" (p. 61, note 60). The reason Strawson wishes to accept this form of dual-aspect monism is that he feels this is necessary in order for pure panpsychism to engage with the known facts of the natural sciences (p. 61).

Premise [2] is paradoxical, but elsewhere Strawson gives an extensive discussion and motivation for it (2006b, pp. 230-43). He uses energy as an example of something that 'from the outside' can only be discussed in non-experiential terms (in the scientific language of kinetic energy, of a spatial distribution of potential energy, and so on); and 'from the inside' must be spoken of (if we are pure panpsychists) in experiential terms. Yet we know that we are speaking of one thing only, namely energy. He admits that the metaphor of 'inside' versus 'outside', being spatial, is misleading (2009b, p. 62). Strawson is fully aware of the paradoxical character of [2]. Consequently, he is prepared to admit that, if ESFD monism is true, and if it is incompatible with fundamental principles of thought and language, then "this is just one more proof of the limitations of human understanding" (2006b, p. 242). This last remark is of course a huge bullet to contemplate biting.

Rejecting ESFD monism

Strawson hopes to marry pure panpsychism with ESFD monism by declaring an identity: 'from the inside' sesmets are centres of experience, but they are identically 'from the outside' "a portion of energy stuff" (2009b: 60-61). Only in this way, he believes, can pure panpsychism be made compatible with established truths of science. He expresses this as, "It cannot be a betrayal of pure panpsychism to require this [marriage], if pure panpsychism as I understand it is to be true, for it must I take it accommodate the existence of such real natural facts as the facts of (say) reproduction and evolution" (p. 61). His use of the phrase "cannot be a betrayal" is a frank acknowledgement of how difficult it is to reconcile these distinctive metaphysical positions within one coherent system.

In my view, there are several reasons for rejecting ESFD monism and holding pure panpsychism alone:

(1) The paradoxical nature of ESFD monism is not sufficiently cleared up by Strawson's example of energy. As with all identity theories there remains an explanatory gap, in this case between energy described in (non-experiential) physical terms, and energy having some sort of description in terms of experiences and volitions. Explicit detail is needed showing how these are related, and in particular, how and

why particular instances of each are to be identified.

(2) There is a contradiction between pure panpsychism and ESFD monism. According to ESFD monism, energy (or any physical entity whatsoever) has, among other things, a non-experiential mode of being. By ESFD [3], this mode of being is neither based on nor realised by nor in any way asymmetrically dependent on energy as conceived experientially. But pure panpsychism, being a form of idealism (mentalism if you prefer), affirms that all (seemingly)-non-experiential being is ontologically grounded in experiential being. So, whereas ESFD monism affirms symmetry, pure panpsychism denies this symmetry. This is a clear contradiction that the obscure character of ESFD monism might camouflage but cannot eliminate: However it is expressed, ESFD monism is plainly intended to hold that there is symmetry between the experiential and the non-experiential.

(3) ESFD monism is superfluous. The bulk of the present chapter has already demonstrated that it is possible to give a full account of the scientific facts of the world – including energy, reproduction, evolution, and cosmology – solely in terms of pure panpsychism (equivalently pan-idealism). What are we relinquishing if we altogether stop giving credence to the concept of 'the physical **as such**'? In a word: Nothing. For example, the notion that there is a 'real physical distance as such' between objects A and B – one which has some meaning in the absence of **all** experience – has never been made intelligible by physicalists. In contrast, according to pan-idealism, everything claimed by scientists to exist indeed does so: The distance between A and B can be expressed in solely experiential terms. Moreover, entities A and B exist in essentially the same way that we do as persons, themselves having percepts and volitions (albeit perhaps much simplified).

Galen Strawson affirms that if ESFD monism is in the end incoherent, then "it is the non-experiential, not the experiential, that must give way" (2006b, p. 243). I agree with this, and believe that retaining ESFD monism is a mistaken attempt to cling on to the vestiges of physicalism, which holds most present-day philosophers in thrall. Once ESFD monism is discarded, pure panpsychism is identical to pan-idealism.

Coming next

Pan-idealism will be developed further. Chapter 9 investigates mental causation and free will. Chapter 10 is a substantive attempt to link pan-idealism to the specific physics of our universe.

Chapter 9

Free will

The chapter begins with a brief account of the currently dominant theory regarding free will – one in which this faculty is compatible with determinism. This is critiqued on the grounds of its implausibility. Then Robert Kane's well-developed libertarian position is presented in some detail. Kane does not claim that he has solved all the problems of libertarianism – he merely asserts that its only difficulties are not additional to, but rather are shared with, those of the compatibilist position. Kane's theory is especially attractive because it fits in with our pre-theoretic notions about the world.

My own contribution is to show how pan-idealism, described previously, enables us to ascribe a strong form of agency to humans (and other beings), in such a way as to remove the above-mentioned difficulties in Kane's position. In the next chapter I will give precise details as to how such free will might be physically realised.

This chapter ends with Erwin Schrödinger's lucid arguments to the effect that the randomness of quantum mechanics is of no help in giving a morally serious account of libertarian free will. For this reason, and despite what he concedes is our "inalienable" intuition that we possess such free will, he eventually settles for a compatibilist position. I reject this conclusion by explaining that it is implicitly founded on mistaken physicalist assumptions.

Compatibilism

In our everyday lives we do not doubt that we have genuine free will. In my current situation I could continue to struggle to write, or I could break off and have a cup of tea for instance. The position of more than 99 per cent of philosophers is that this choice depends on the fine details of my brain's physical state during the brief interval in which the 'decision' is made: Because my brain was in a particular (type of) physical state, I have continued writing; Had my brain been in another type of state I would have stopped for tea; In yet another type of state I would have gone for a walk; And so on. This majority position concerning the character of free will is called *compatibilism* because it ascribes my 'free choice' to the details of my brain's physical state, and the physical processes going on within it, during the period of decision; and so it is necessarily compatible with the laws of physics (which are either deterministic or random).

There are several problems with this position. First: What relevance do these counterfactual versions of me (with slightly different brain states and processes) have to my so-called 'choice'? There is only one physical me, with one actual current brain state; thus my so-called choice is fully determined (if it is not random). Second: What counterfactual variants of my brain are to be counted as being 'valid alternates that are still essentially me'? And on what rational basis can this be decided? Is the hypothetical removal or addition of a tiny portion of brain matter okay? How extensive could any hypothetical rewiring of my neurons be, while still counting as a valid variant of my brain? Presumably such imagined near-duplicates would superficially look indistinguishable from me. But, by definition, such counterfactual variants **do not exist**. Under compatibilism my 'choice' is not real.

Third: Mind has no relevance whatsoever in the compatibilist account of freedom. All that is actually and ultimately going on is physical brain processes causing bodily actions. Hard-nosed physicalists would claim that this is an advantage of their position. For those of us who accept that there are entities in our world possessing irreducible mental properties (as argued in Part I), these properties must have causal effects: nothing real can be an epiphenomenon. In simple terms, my feeling of mild thirst and my strong desire to make progress with this chapter were causally relevant to the choice I made to continue writing.

Robert Kane on libertarian free will

Robert Kane, despite being very much in the minority as a defender of authentic (technically *libertarian*) *free will*, has a high reputation among his academic colleagues. He is for example the editor of *The Oxford Handbook of Free Will* (2002b); and his own position, set out in *The Significance of Free Will* (1998), has received wide praise. It is by far the best developed of all libertarian positions, and has its roots in the ideas of Aristotle.

Kane asserts that for humans to possess libertarian free will it is necessary for them to have the power and ability to perform several alternative actions in a given precisely defined physical situation. This ability Kane calls *Alternative Possibilities* (1998, p. 33). In the example of me deciding whether to continue to write or to have a cup of tea, the physical situation includes not only general facts, for instance that I am sitting at my desk writing, but include everything possible that might be said (in principle) about the physical state of every atom or particle in the room in which I am working. Crucially, to have Alternative Possibilities means that it must be within my power to freely select and act upon any one of them of my own choosing – even given a maximally-exact description of my present brain state. The choice of refraining from doing anything is also considered to be an Alternative Possibility.

The necessity for Alternative Possibilities is generally recognised by all proponents of libertarian free will. It is easy to see that libertarian free will is incompatible with our universe being deterministic. This does not trouble libertarians because the empirical evidence (particularly quantum theory) suggests that physical determinism is false. Opponents of libertarianism go on to argue that if the universe is not deterministic then it must be random, and randomness (they claim) is inconsistent with libertarian free will. This argument will be discussed later and shown to be faulty. But Kane gives a partial answer...

Suppose that as I am now writing I am considering whether or not I will break off to have a cup of tea. My brain is presently in some well-defined but unknown state. There is excellent scientific evidence suggesting that there is some unknowable but definite probability (call it p) that I will break off for tea. How can this decision be considered free when it happens with this definite probability p?

Some other libertarians have tried to evade this conclusion. Niels Bohr, for example, argued that even the most perfect possible description of an object as complex as the human brain cannot involve definite probabilities – **even in principle** (quoted in Schrödinger, 1951, pp. 168-71); and he uses this to defend human freedom. In the same passage, Schrödinger strongly criticises Bohr's position. To give another example, David Hodgson suggests that, because each situation of physical choice is necessarily unique (again for reasons of complexity), probabilities are no longer relevant (Hodgson, 1991, pp. 392-3). This argument is not valid because there are some conceptions of probability that do not rely on repetition. For example, *propensity probability* is defined as the tendency of some experiment or event to yield a certain outcome, even if it is performed only once.

To his credit, Kane doesn't evade this issue: he accepts that there must be a fixed but unknowable probability p that I would have broken off for tea. Given the probabilistic nature of the decision, how can it be a free choice? Kane's answer is that, by choices made over the course of our earlier life, we have chosen p. In detail: We have *Ultimate Responsibility* (1998, p. 35) for our present actions because they are based upon the historical sequence of genuinely free acts in our past – ones over which we have reflected. Kane calls such reflected-upon, conscious free acts *Self-Forming Actions* (1998, p. 75). We are responsible for our present characters because they are built upon our past Self-Forming Actions. Many of our activities are based on habit rather than conscious choice. For most of the time I continue to write without any thought of stopping. It is a well-established psychological fact that, as I continue to make conscious decisions, I become more likely to make similar decisions in the future. Thus, the probability that I will take a tea break is based on my past history, and in particular on my past conscious choices. The valuable concepts of Ultimate Responsibility and Self-Forming Actions are original to Kane.

The ways in which we mould ourselves by our Self-Forming Actions are much more significant than the writing versus tea drinking example above. Discussion of libertarian free will often centres importantly on our development as ethical beings, and our role in this. The development of our characters in terms of charitableness and honesty gives two examples of this. For most of the time I may act through habit, but there will be occasions in my life where I will have to make tough, costly, conscious choices, and these will help to form who I am as a moral being. It is crucial to understand that this theory implies that if a person's upbringing and environment are impoverished then the extent to which they possess libertarian free will is correspondingly diminished (or in extreme circumstances even eliminated). Robert Kane expresses the point thus:

> Another kind of disservice is done by continuing to appeal to a mysterious form of agency to account for free will. [...] This can easily lead one to think in turn that the abused child or ghetto dweller has as much free will and ultimate responsibility for what he or she does as one who lives in more advantaged circumstances. [...] The theory of this book implies no such consequence. Precisely because that theory recognises the embeddedness of free will in the natural order, it recognises that free will and moral responsibility

> are a matter of degree, and our possession of them can be very much influenced by circumstances. That is why, if one believes in the value of free will and ultimate responsibility, it is important to cultivate a social order in which they can flourish...
> (Kane, 1998, p. 213)

Far from implying a punitive, socially conservative position, one in which even a lifetime of extreme abuse is never a mitigating circumstance for an act of wrongdoing, Kane's position implies that we have a political duty to encourage a social environment in which every person has the opportunity to flourish.

Weighing up alternatives

An analogy has entered our language likening making free choices and the process of weighing articles. But it is important also to remember the disanalogies. In simple cases the analogy holds quite well. I enter a grocer's to buy an orange. I am likely to choose the biggest on display, and I might even pick two up – one in each hand – the better to compare their weight.

But suppose I want to buy a car. There are many factors that I will want to take into account: The price; running costs; environmental impact; reliability; durability; maximum speed; aesthetics; brand preference; manufacturer's ethics (workers' conditions); and so on. This is a multi-factorial problem. Some factors are numerical and well-defined (e.g., cost); others are numerical but must be estimated (e.g., reliability); some are not numerical (e.g., aesthetics); and so on.

These factors cannot be directly compared to one another – it is, at least in part, up to me which factors I give more weight to. I might be willing to pay more for a car based on its reputation for reliability. The point is that I can give a rational argument for my choice of car based on the weights I give to different factors. Another person, who assigns different weights to these factors, can give rational reasons for choosing a highly contrasting car. This freedom to choose which weights to assign, leading to contrasting rational choices, is an important disanalogy with the literal act of weighing, in which Nature assigns the weights (see Hodgson, 1991, chapter 5).

Wanton freedom

If libertarian free will exists, it is possessed only by humans and perhaps, to an extremely limited extent, by others of the most complex social animals, such as hominids. Moreover, it is only possessed by such individuals as they progress beyond infancy towards adulthood and a significant number of Self-Forming Actions have occurred. Kane's theory makes clear that libertarian free will depends upon memory of past events, knowing the likely consequences of personal actions for events in the future, and the rational weighing of options.

Babies do not possess libertarian free will in Kane's sense, nor do they have sufficient rationality or experience to properly weigh decisions. What do they have? Kane calls the kind of freedom that babies have *Wanton Freedom*. This is (initially) Alternative Possibilities alone. As a baby begins to develop, she randomly strives to act, and some of these volitions happen to move her limbs at random. By repetition she learns the consequences of these initially-random volitional acts, and by this means she gradually brings her limbs under control. In the same way she learns to pick up and drop objects – the extensive repetition needed to learn this skill often tests the patience of parents. (This is not to deny that some skills, such as the ability to walk, use language, or recognise faces, are genetically encoded to a certain extent.)

Except in disastrous circumstances such as Kane describes, by the age of about four years children can empathise with others, know which of their everyday actions are right and wrong, and make simple moral decisions. They are also just beginning to learn how to keep their emotions under sufficient control to be good friends with their playmates. At this stage of their social development they make many naive mistakes. For this reason, it is almost universally accepted that people are only regarded as being criminally responsible for their actions once they have reached a certain age. This is true worldwide, despite the fact that the age of criminal responsibility varies considerably (between about 7 and 21 years) for different crimes and for different cultures. In the United States, an illogical anomaly regarding the age of responsibility there, is that children below it can be "tried as adults" if their alleged crimes are sufficiently serious.

So, a baby initially possessing only Wanton Freedom develops into an adult with libertarian Free Will. Does this present any problems either physically or philosophically? I would say **No**: Libertarian free will requires the ability to weigh options, foresee consequences, and to take appropriate actions. All of these can be encoded in the increasingly complex neural structure of the individual as they grow to adulthood. Likewise, in principle, there is no problem in a species having only Wanton Freedom evolving into a more sophisticated one possessing libertarian free will.

Critique of Kane

Kane's theory does not, of itself, solve the mind-body problem

Kane's theory of libertarian free will is far more complex than has been described here. He puts forward ninety-one theses concerning his main postulates, with extensive commentary upon each. Thesis 51, in which he asks, "How can thoughts, perceptions, and other conscious experiences – including efforts of will – **be** brain processes?" (p. 148, my emphasis), makes it clear that he is a physicalist, and in particular favours some form of identity theory. He admits that the mind-body problem remains mysterious, but he argues that such difficulties afflict compatibilists just as much as libertarians.

Are Alternative Possibilities consistent with physicalism?

Kane maintains his physicalist position consistently. He is therefore adamant that libertarians must reject any form of causation by agents that is either different from or over-and-above physical causation. He calls such purported, illegitimate power to choose *agent-causation* (hyphenated) in contrast to legitimate agent causation (unhyphenated) that is somehow consistent with physical causation. He criticises agent-causation extensively throughout his book (in particular, see pp. 120-3 & 187-95).

But – at least on the face of it – Kane's (in my view solidly grounded) declaration that humans possess Alternative Possibilities **seems** to assert that we possess a kind of mind-body causation that contradicts physicalism. Recall that Alternative Possibilities means that it is within my power to freely select and act upon any one of several bodily actions of my own choosing – even given a maximally-exact description of my present brain state. Alternative Possibilities are definitely a type of causal power. Moreover, they seem to be inconsistent with the mathematical character of all causation in any physicalist universe: Even if the relevant physical laws are probabilistic, these laws will be to the effect that my future brain state occurs at random, rather than being the result of a free choice.

Kane gives his most explicit account of Alternative Possibilities in terms of physics in his Theses 25 and 26 (p. 130). Here he identifies efforts of will with indeterminate physical processes in the brain that are chaotic and sensitive to quantum indeterminacies. He then sets down a hypothetical objection, which he addresses to himself:

If these indeterminate efforts truly are macro indeterminate

> processes in the brain, as you [Kane] suggest, it is difficult to see how the choice outcomes that result from them could be in the agent's control rather than merely matters of chance. (p. 131)

Kane's response to this objection begins:

> The first thing we have to remind ourselves here is that the indeterminate processes in the brain [described in these two Theses] are *also* physical realisations of the agents' efforts of will and are experienced by the agents as something they are *doing* – that is by struggling to resolve inner conflicts...
> (p. 131, original emphasis)

In this response, Kane has moved towards a double-aspect monist theory of mind and body, in which these two latter things are identical, but where mental descriptions cannot be reduced to physical descriptions (or vice versa). Both viewpoints have equal weight: a person struggling to make a moral decision and a certain type of physical brain process are one and the same. A severe difficulty of such a (non-panpsychist) double-aspect monism is that the mental aspect is supposed to be confined to an extremely small region of spacetime, namely that within brains. Moreover, this mental aspect is of an entirely novel character – one that is inexplicable in principle from the most perfect possible description of the brain as a physical system. Prior to the evolution of sophisticated brains, our universe existed for billions of years devoid of experience. Theories of this type assert that consciousness is a radically emergent novelty; and such positions have been criticised in Part I.

How do Self-Forming Actions get started?

Another criticism concerns the important link between free will and ethics. Kane's position is that an adult's current propensities to act well or badly are grounded on their personal history of Self-Forming Actions. Our present thought process is conditioned by the laws of physics and our current brain state, but this state was brought about by these Self-Forming Actions. Our freedom lies in our personal history. This is certainly true as far as it goes, but a difficulty remains unresolved.

Each person must have had a **first** Self-Forming Action. This would have been as a young child, with extremely basic notions of right and wrong, and equally primitive notions of the consequences of their actions. The decision might involve whether or not to eat a sweet belonging to a friend; or it could be something entirely different. Whatever the details, Kane's theory requires that an initial Self-Forming Action exists. But now this decision is not grounded in previous Self-Forming Actions – it can only be based on the child's physical brain state, on the laws of physics and on chance. Hence the sequence of Self-Forming Actions cannot get under way.

Pan-idealism and libertarian free will

Kane's theory becomes immune to the above criticisms if one takes a certain philosophical step: that of rejecting physicalism, and instead adopting the position of pan-idealism. Recall that pan-idealism asserts (among other things) that: The only entities that exist in the universe are centres of experience, called *experients*. Their experiences include percepts of other experients. They have a certain lawful, volitional freedom to change their percepts. This wilful freedom is the only form of causation in the universe. The percepts of the experients can be correlated (matched) with one another to varying extents. The claim that some adequate matching exists is identically the claim that the experients coexist in the same universe – and also that they indeed perceive one another. Included in the matching are structural features of the experiences, and we accept the maximally correlated combined structure of all the experients to be, **by definition**, the *physical structure* of the world. The basic entities of (objective) physics – which scientists attempt to discover through their combined experiences – are in fact experients. Pan-idealism is a form of idealism because physical states, causes, entities and laws can be reduced without remainder to mind-like terms. It is also fully realistic about the entities found in the world.

Pan-idealism denies causal powers to any physical states, S, or to any physical laws, L – even those of objective physics. Physical states and laws are nothing other than formal mathematical descriptions, that maximally correlate – across all experients – structures that exist within their percepts. Thus, agent-causation (**hyphenated**: which Kane defines as causation that is (a) non-physical, and (b) specifically **by agents**) cannot be in conflict with physical causation, because the latter simply does not exist within pan-idealism.

As already discussed, very young babies have Alternative Possibilities, which Kane calls Wanton Freedom (in contrast to libertarian free will which additionally requires substantial cognitive powers). The terms 'agent-causation', 'Wanton Freedom', and 'Alternative Possibilities' can be regarded as synonymous in pan-idealism.

There is no problem in assuming that all experients in our universe have Wanton Freedom, and that this is the only form of causation. Photons have this Wanton Freedom for example. If photons emerge from a source, pass through a very narrow slit and land on a photographic plate, then where each individual photon lands cannot be predicted: only statistical predictions about a whole group of photons can be made. But "Why did this particular photon land on this particular spot on the plate?" is a question that calls out for explanation. The answer provided by pan-idealism is that this location is the Wanton Free choice made by the photon. The statistical results *arise* from the propensities of the photons to make various free choices in the given situation.

How can a baby possessing only Wanton Freedom grow into an adult with libertarian free will? A baby, brain included, is an incredibly complex hierarchical system of experients. The development of free will involves rationality, the ability to weigh alternatives as sketched above, memories of the past, and the ability to predict the likely effects of one's actions. It requires self-awareness and empathy with the feelings of others. I do not have a detailed theory, but I am assuming that nothing essentially different from standard cognitive science is needed. The pan-idealist accounts of memory and imagined futures, for example, will be that they are encoded within the structures of the brain. The explanation will be strongly analogous to a physicalist, cognitive account, at least in terms of structure and functionality; but the understanding, of what the brain and its constituents **actually are in their essential natures**, will be vastly different.

There is a huge difference in complexity between the (rational, reflective) consciousness of an adult human and the ('raw') percepts of a simple entity such as a photon. For this reason, some panpsychists prefer to reserve the term 'consciousness' for humans and higher animals; and to use a different term, 'experience', for the much more primitive consciousness of everything else. I prefer not to make this distinction because, as outlined in the previous paragraph, I do not believe it to be a fundamental one. Even the minuscule consciousness of a photon implies an extremely primitive self-consciousness. In its percept, the photon must locate itself, at least approximately, with respect to the experimental set up, or to the natural environment in which it finds itself.

Finally, even the first Self-Forming Action of an infant is, in its fundamental nature, a (primitive) act of conscious will. The physical consequences (including the unknowable probability p associated with the act) are all secondary. Thus, the sequence of Self-Forming Actions can validly get under way.

Arguments against libertarian free will

Compatibilists have made several standard arguments against the possibility of libertarian free will in order to bolster their own position which, as we saw, is on the face of it implausible. Here I want to list these arguments and show that they all fail. The first three are closely related and have already been discussed; they all depend on the presupposition of physicalism. These are:

Objection 1: *Alternative Possibilities cannot be explained in terms of physical randomness.* Recall that all libertarians agree that Alternative Possibilities (at least in humans) are essential to their position. However, if one rejects physicalism then, as we have seen, there is no longer any need to explain anything in physical terms, and so the objection fails.

Kane's own response is that he is a double-aspect monist – a position very close to physicalism – and he argues for the existence of Alternative Possibilities on this basis. He does not claim that this explanation is completely satisfactory; but he merely affirms that this position on the mind-body problem raises no extra problems additional to those faced by all compatibilists. In contrast, I have argued that pan-idealism solves this problem completely because it entirely rejects both physicalism and the concept of physical causation. In line with this, pan-idealism denies that randomness amounts to an albeit-weak form of physical causation. In pan-idealism, it is physical randomness (including probabilities) that is to be explained in mentalistic terms.

Objection 2: *Alternative Possibilities cannot exist because of the causal closure of the physical world.* By definition, the *causal closure of the physical world* is the principle that all physical effects must have solely physical causes. This principle implies that there can be no mind-to-body causation (for example, my conscious urge to raise my arm cannot cause it to move). According to our commonplace intuitions, the claim that we cannot move our own bodies under our conscious control seems implausible, to say the least. But, **under the assumption of physicalism, causal closure is inevitable;** and moreover, it seems natural if one holds the view, common among scientists, that human beings are, in their most basic essence, complex physical systems, subject to physical laws.

If one rejects physicalism, however, there is no need to assume causal closure. Indeed, according to pan-idealism (which does not reject any empirical scientific result) there is no physical causation.

Objection 3: *Agent-causation* (hyphenated) *is a spooky addition to the natural physical causation of the universe.* As we have seen, Kane develops his libertarian theory without agent-causation. If one rejects physicalism then one can deny the existence of physical causation and take agent-causation as the sole form of causation in the universe: agent-causation is basic and is not additional to anything.

The above three objections are all very closely related. Indeed, they can be regarded as three distinct ways in which compatibilists have framed the same argument.

Objection 4: *Libertarian free will is 'immoral' because it necessitates a punitive attitude towards individuals regardless of their circumstances.* Simplistic libertarian theories may have this fault, but Kane's position does not, as is made clear in his description given above about "the abused child or ghetto dweller". A socially conservative punitive attitude, which might be expressed as, "Factors such as impoverishment or an abusive childhood are never mitigating circumstances for bad behaviour, because many others who grow up in such environments go on to lead impeccable lives", is incorrect. First, it is factually wrong: there is a strong correlation between such appalling circumstances and the likelihood of moral failure. Second, Kane's theory depends on the full particulars of an individual's life history. Whether or not an individual comes into some general category, such as having below £X to live on, or was referred to social services because they were abused, does not amount to anything approaching the full account of the life-history of that person.

Kane's theory suggests that a well brought up person of previous good character who commits a certain crime is more culpable than a highly-disadvantaged morally-damaged individual who commits a similar offence. Unfortunately, in many jurisdictions, the punishments handed down by the courts often seem to reflect the opposite view.

Ultimate Responsibility for one's actions is a desirable thing, and something most people would want to have. First, not to have it would rightly be considered to be infantilising. Second, not to have it (or to have only limited Responsibility) is a sign of an extremely damaging personal history, and no one capable of reflection would wish that upon themselves.

Objection 5: *The possibility that so-called 'Frankfurt controllers' might exist shows that the libertarian conception of free will is wrong.* In recent years, philosopher Harry Frankfurt has put forward a very influential, argument against libertarian free will that has received considerable discussion (see Kane, 2002b, Part V). It claims to show that Alternative Possibilities are irrelevant as to whether or not a person possesses free will. Frankfurt proposes the following thought-experiment: The sinister Green wants to control Jones into performing a certain action. He keeps a covert watch on Jones as he comes to his decision. If Jones is about to decide to carry out the action, then Green does nothing; but if Jones is about to decide the opposite, then Green steps in to ensure that Jones carries out Green's will. Suppose on this particular occasion, Jones decides to carry out the action and does so. Because Green has done nothing to influence Jones, it would seem that Jones has acted freely, even though it would not be possible for him to do otherwise. More recent variants of the thought-experiment have replaced Green with an automatic monitoring device (say a microchip, 'auto-Green') implanted in Jones' head, but working in essentially the same covert way. Entities such as Green and auto-Green are called *Frankfurt controllers*.

Kane's response to Frankfurt is in two parts. First, unlike more simplistic libertarian positions, Kane's theory does not assume that every attempt by Jones to exercise free will needs to be successful, only that in the course of his life Jones is successful on a significant number of occasions. This means that the auto-Green device would have to be implanted before birth, and remain in place throughout Jones' lifetime up to the present moment. Moreover, the device would have to work near-perfectly. This puts severe constraints on Frankfurt controllers, and somewhat undermines their conceivability. Kane's second point is more telling (his Thesis 41, p. 142). Acts of willing could be indeterminate, such that it is impossible to tell what decision is made until after it happens. This denies that Frankfurt controllers are possible, even in principle, and so Frankfurt's argument fails. A weakness for Kane here is that he does not describe in terms of physics how these acts of willing are to be realised within the brain – for it is the specific detail of this physical implementation that governs whether acts of willing are indeterminable or not.

The next chapter will give my specific realisation of pan-idealist free will in terms of quantum physics. This is done in such a way that the times of acts of willing are impossible to determine before they have occurred. If pan-idealism and its physical realisation are correct, then Frankfurt controllers are impossible in our universe. (If these ideas are wrong, then at least I have described a *conceivable* universe in which libertarian free will holds and Frankfurt controllers are impossible, thus opening up the possibility that our world has these characteristics.)

Objection 6: *Neuroscience experiments prove that libertarian free will is impossible.* In the 1980s, Benjamin Libet performed a series of experiments apparently showing that brain activity initiating a voluntary action (a hand movement) commenced about half a second before the subjects made the conscious decision to perform the action. Some claimed this showed that free will is an illusion because the conscious decision came too late: the brain, as an unconscious physical system, had already initiated the movement. The original experiments had limitations, and it is better to discuss a later one by Chun Siong Soon and others (2008):

> The subjects were asked to relax while fixating on the center of the screen where a stream of letters was presented [one at a time, at half-second intervals]. At some point, when they felt the urge to do so, they were freely to decide between one of the two buttons, operated by the left and right index fingers, and press it immediately. In parallel they should remember the letter presented when their motor decision was consciously made.
> (2008, p. 543)

By monitoring each subject using an fMRI scanner, the experimenters were, in effect, able to predict, with 95% accuracy, which button they were going to press, 7 seconds before the subject's **self-reported** time of their decision (given by the remembered letter). Because of a 3 second delay within the operation of the fMRI scanner, the effective predictions were 10 seconds in advance of the subjects' reported decisions (pp. 543-4).

There are two responses to this admittedly remarkable result. First, the experiment involves subjects performing immediate, insignificant, quasi-random actions at quasi-random times. It is not clear that the same results carry over to ethically significant and rationally considered actions carried out at appropriate times, and after due thought. (Some ethical situations were described in the account of Kane's theory; and Schrödinger provides further examples in the next section.)

Second, the experiments assume that subjects have some insight as to the moment at which their conscious decisions are reached. There is no reason to believe that this is so. Suppose humans possess libertarian free will. There is no evolutionary advantage in their developing the ability to record the moment each conscious decision is made: indeed, in situations of danger, such recording would be a harmful distraction.

My hypothesis is that, having pressed the button, the subject remembers the last few letters presented, and confabulates one of these as being the likely time of their decision. It may seem implausible that human insight, as to the moment when we make decisions, can be up to ten seconds out. In my view, however, this is far more plausible than the bizarre assumption that "conscious decisions to act" – which everyone **knows** they make every day – are a wholly ineffectual delusion.

Schrödinger on free will

Erwin Schrödinger has given a lucid account, in which he reluctantly but definitively comes down in favour of compatibilist rather than libertarian free will (2014/1951, pp. 162-8). His discussion is entitled "Would physical indeterminacy give free will a chance?" I will summarise his arguments here, showing that they depend upon an unquestioned presumption of physicalism.

He first gives full weight to our intuition that we possess libertarian free will, calling it "inalienable":

> [There is the] apparent contradiction between the deterministic view about material events and what is called [...] in modern language free will. I suppose you all know what I mean: since my mental life is obviously bound up with the physiological goings on in my body, more especially in my brain, then, **if the latter are strictly and uniquely determined by physical and chemical natural laws**, what

> about my inalienable feeling that *I* take decisions to act in this or that way, what about my feeling responsibility for the decision I actually do take? Is not everything determined in advance by the material state of affairs in my brain, including modifications caused by external bodies, and is not my feeling of liberty and responsibility deceptive?
> (Italics original, bold emphasis added, pp. 162-3)

The phrase in bold is where the assumption of physicalism is made. He goes on to call this conflict "a true *aporia*", i.e., a true contradiction or paradox. In the above passage he is referring to the strictly deterministic world of classical physics. He goes on to note that quantum physics is indeterministic, and then asks:

> Could perhaps the declared *indeterminacy* allow *free will* into the gap in the way that *free will determines* those events which the Law of Nature leaves undetermined? This hope is, at first sight, obvious and understandable.
> In this crude form the attempt was made, and the idea, to a certain extent, worked out by the German physicist Pascual Jordan. I believe it to be both physically and morally an impossible solution.
> (Italics original, p. 164)

Schrödinger may have been right about Jordan's position, but in my view this general approach can be developed acceptably – provided one adopts Kane's libertarian position together with my pan-idealism.

Schrödinger offers two physical objections. The first is that acts of will (what Kane calls agent-causation) would necessarily involve interference with, and violations of, the statistical laws of quantum mechanics (pp. 164-5). According to pan-idealism, however, the **sole** causation in the universe **consists of** agent-causation (mostly by primitive experients having wanton freedom alone – free will is the preserve of some complex experients). Physical goings-on (including the **standard and undisrupted** statistical laws of quantum mechanics) are **constituted by** these acts of agent-causation – which include acts of libertarian free will. His second physical objection is that:

> Now we know that *there is no statistics in the reaction of the same person* to precisely the same moral situation – the rule is that the same individual in the same situation acts again precisely in the same manner.
> (Italics original, p. 165)

Kane's theory of Self-Forming Actions explains how people on occasion react differently in exactly the same moral situation; and how this develops their moral characters. He also explains how they act by habit "as a rule". Physically identical brains can act in different ways in identical situations because they are quantum systems. This has been most explicitly explained by Stapp (2007, chapters 4-6). Schrödinger is wrong in his last assertion.

Here is Schrödinger's moral objection:

> The moral objection was strongly emphasised by the German philosopher Ernst Cassirer (who died in 1945 in exile from Nazi Germany). Cassirer's extended criticism of Jordan's ideas is based on a thorough familiarity with the situation in physics. [...] Free will in man includes as its most relevant part man's ethical behaviour. Supposing the physical events in space and time actually are to a large extent not physically determined but **subject to pure chance**, as most physicists in our time believe, then this **haphazard** side of the goings-on in the material world is certainly (says Cassirer) *the very last to be evoked as the physical correlate of man's ethical behaviour*. For this is anything but haphazard, it is intensely determined

> by motives ranging from the lowest to the most sublime sort, from greed and spite to genuine love of the fellow creatures or sincere religious devotion.
> (Italics original, bold emphasis added, p. 166)

Here again, the passages in bold indicate unquestioned physicalist assumptions. Rather than being "subject to pure chance" or "haphazard", the stochastic character of physical laws is grounded in acts of volition. As already explained, with certain complex experients, these acts of volition can amount to conscious acts of libertarian free will. Moreover, they can do so in such a way as to be (as Kane explains) a robust and far from haphazard ground for ethics.

Schrödinger, not considering alternatives to physicalism, comes down in favour of compatibilist free will. I hope the arguments given here show that, if one rejects physicalism, then libertarian free will remains an attractive option, not least because it is in line with our inalienable intuitions about ourselves. Of course, we must ensure that any new metaphysical system we adopt, in this case pan-idealism, is consistent with modern physics: this will be the next topic.

Chapter 10

Our pan-idealist universe

This chapter presents my pan-idealist model, which crucially gives a rational account of *mental causation*, including free will and agency in humans – and more basic agency in simpler experients. Another significant feature is that it provides a natural explanation of the manner in which centres of experience unite. More explicitly, the *combination problem* asks: How do physical ultimates having primitive mental properties combine to form unified minds of the kind that we know ourselves to possess? An excellent solution is provided here to what many still regard as an insuperable problem.

Recall from chapter 6 that the Ghirardi, Rimini and Weber (GRW) interpretation of quantum mechanics has many attractive features. It naturalises measurement, gives a unified account of the macroscopic and the microscopic, and it also explains why macroscopic objects have well-defined trajectories. A summary of standard GRW is given here, and further detail regarding measurement in GRW will be presented.

Pan-idealist GRW makes use of the mathematical formalism of standard GRW, but adapts its meaning by instead proposing that our universe consists of nothing else besides agents possessing free will in the most basic (Alternative Possibilities) sense of Kane (chapter 9). Localisations depend on the choices of agents, whose propensities to act are given by the GRW statistical rules. The model has the following features

- It specifies the agents in explicit physical terms
- It explains how inconsistencies among agents' choices are resolved
- It explains how the universe's fixed past develops over

time by adding single layers of spacelike separated localisation events
- It gives an account of how agents come into being, combine and die.

Despite these philosophical differences this model is empirically (i.e., in terms of any scientific experiment) identical to the original GRW model.

Recap of pan-idealism

Here is a summary of pan-idealism, using the postulates **Pa** through **Pi** listed in chapter 8.

Pan-idealism is a form of idealism: The universe, at its most basic level, consists of nothing else than centres of experience, called *experients*, that can perceive (at least some) other experients. The only form of causation in the universe is the free agency of these experients: By an act of volition, an experient can change its own percept, and also the percepts of the experients that perceive it (**Pe** through **Pg**). Most experients have zero cognitive powers even when, as frequently happens, they exist in hierarchies. Hierarchies can occasionally form themselves into organic individuals, a limited number of which possess cognitive powers (**Ph**).

The 'pan' of pan-idealism: Pan-idealists are realists because they hold that the universe has an objective (true) physics (**Pa**). Moreover, pan-idealists agree with physicists about the physical laws that most accurately describe the universe – given the best current knowledge. As was made clear in chapter 8, the essential reason for this agreement is that scientific practice is entirely based on human experiences.

Pan-idealists are in almost complete accord with physicalists about the contents of the universe (**Pb** through **Pd**). (Pan-idealists do make a distinction between a *true individual*, which is an experient in its own right, and a mere *aggregate* of such individuals, which is not of itself an experient. Most physicalists regard this distinction as meaningless.)

The mind-body problem for pan-idealism is to demonstrate how the entirety of objective physics can be derived from mental facts alone (**Pi**). A large portion of chapter 8 was devoted to showing how this must be possible – at least in principle – no matter what the objective physics of our universe happens to be. A key idea is to let physical properties be characterised as being **identically** the *maximal intersubjective consilience* between specific structures existing within the percepts of experients.

Recap of GRW

This section recalls some salient points of standard GRW theory, which was detailed in the **Ghirardi, Rimini and Weber** section of chapter 6. In GRW, the wavefunction is preferentially expressed in terms of position variables:

$$\Psi(\mathbf{q}_1, \mathbf{q}_2, \ldots, \mathbf{q}_N; t) \tag{1}$$

Here $\mathbf{q}_k = (x_k, y_k, z_k)$ are the positional coordinates of the kth particle. For most of the time, particles obey the Schrödinger equation. Occasionally, however, a particle, again say the kth, will localise its position so that it is very close to a specific place, say $\mathbf{q}^0$.

Each particle is as likely to localise as any other. A given particle localises at random times (these occurrences have a Poisson distribution). This happens extremely rarely, with the average time, τ, between localisations being once in 10^{15} seconds (about once every 100 million years).

A localisation is somewhat akin to an approximate position measurement. But it is described entirely in objective physical terms, and without reference to any observer (whether conscious or not). The inexactness of the measurement is given by parameter $\sigma = 10^{-5}$ cm. The physical significance of σ is that, with probability of about $^2/_3$, the given particle will localise within a distance of σ of $\mathbf{q}^0$. Because localisations are akin to position measurements, they follow the Born Rule; localisations tend to occur where the wavefunction Ψ predicts them to be.

The centres of localisation, such as $\mathbf{q}^0$, are points in spacetime that give physical meaning to GRW theory, and are called *flashes*. Spacetime is peppered with such flashes, and the corresponding ontology is called *flashy GRW*. There is a slightly more complicated *massy GRW* ontology, which was explained in chapter 6. I will still call the central peaks, $\mathbf{q}^0$, of mass localisations *flashes*, even though this terminology is non-standard in massy GRW. I am not particularly concerned about the distinction between the flashy and massy ontologies.

In GRW only the wavefunction Ψ, given by (1) above, and the flashes are real. A particle isn't a thing that exists separately: it is merely a triplet of arguments from Ψ, such as $\mathbf{q}_k = (x_k, y_k, z_k)$. (We are ignoring spin for the moment.) Nonetheless, it is often useful to speak of particles, and I'll continue to do so.

Measurement in GRW

As we saw in chapter 6 most interpretations of quantum mechanics do not give an adequate account of measurement: What constitutes a measuring device? Is consciousness required for a measurement? And so on. Here I will give an explicit and objective account of measurement, as this is explained in GRW theory.

Measuring devices

Figure 10.1 sketches a device to measure the z-spin of a spin-½ particle, $|\mathbf{s}\rangle$. The device has a pointer that moves in a slot which has labels: **Down**, **Start**, and **Up**. The device is also labelled with an arrow marked **z**, to help the user orient it correctly. We may suppose the device contains a Stern-Gerlach apparatus within it – compare this sketch with Figure 4.3. However, here we are treating measuring devices as black boxes rather than being concerned with their inner workings.

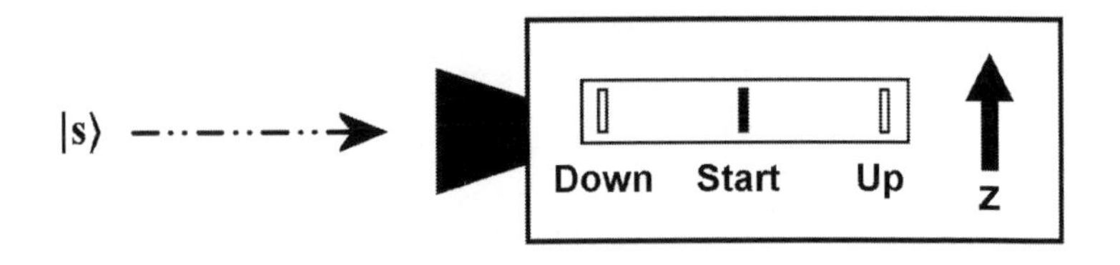

Figure 10.1: A measuring device to measure the z-spin of a spin-½ particle, |s⟩

Before each experiment, the pointer is reset to the **Start** position. The particle enters the receptor at the left. The device has recorded the spin of the particle as being either **Down** or **Up** once the pointer has moved to one of these respective positions. No other outcome is possible. In order to be a useful measuring device, it must have the following properties:

1. If $|\mathbf{s}\rangle = |\mathbf{z}\uparrow\rangle$, then the pointer invariably moves to the **Up** location
2. If $|\mathbf{s}\rangle = |\mathbf{z}\downarrow\rangle$, then the pointer invariably moves to the **Down** location

These situations would arise if the particle had been prepared appropriately. Alternatively, such a particle, arriving unprepared from the surrounding natural environment, might be in one of these two states by chance. Moreover, in order to be a valid measuring device, agreeing with the experiment, it must have an additional property:

3. If particle $|\mathbf{s}\rangle$ spins about an arbitrary axis, i.e., if $|\mathbf{s}\rangle = c|\mathbf{z}\uparrow\rangle + d|\mathbf{z}\downarrow\rangle$, with c and d being any complex numbers such that $|c|^2 + |d|^2 = 1$, then the pointer will move to

> the **Up** location with probability $|c|^2$, and to the **Down** location with probability $|d|^2$

See the section **Explaining basic spin experiments/*Spinors and the Stern-Gerlach Apparatus*** of chapter 4. Rule 3 implies the first two rules just given. Because rule 3 is statistical, it can only be checked by preparing many particles with the same initial state $|\mathbf{s}\rangle$ and repeating the experiment many times.

More generally, a *measuring device* is a large physical object that gives a permanent record of the state of a quantum system by say moving a pointer to a different location, or perhaps by making marks on a paper tape (here it is a quantity of ink that changes location). Regarded as a black box, i.e., ignoring the specifics of its interior workings, a measuring device can be described adequately in classical terms. Measuring devices can readily be constructed for determining properties that have more than two possible outcomes.

Aside: Measurement in non-GRW interpretations

Could a measuring device **itself** get into a superposed state such that the recorded measurement was the superposition $c|\mathbf{Up}\rangle + d|\mathbf{Down}\rangle$? Different interpretations of quantum mechanics give different answers. The following can be regarded as summarising chapter 6:

Within the **Copenhagen School**, there was, and still is to this day, no agreement. On different occasions Bohr himself said that the question could not even be asked; on others he claimed that environmental decoherence meant that the device would – of itself – quickly collapse into one of the two pointer positions. Heisenberg always insisted on the latter environmental collapse, so that the device would quickly decohere into either **Up** or **Down** – even in the absence of any observer. In contrast, Peierls insisted that a conscious observer is essential for collapse. For him, in the absence of a conscious observer the device would remain in a superposition of states indefinitely. Von Neumann held that the task of physics is merely to give a theory that correctly predicts human observations: what is really going on is irrelevant.

According to the **pilot-wave theory**, although the wavefunction can exist in superposition, the particles themselves move along exact (but not precisely knowable) trajectories, under the guidance of the wavefunction. The theory is such that all of the particles within the pointer move in unison, either to the left or right (in a manner consistent with the three rules above). The measuring device, regarded as a black box (i.e., as a system of particles, ignoring the wavefunction), never exists in a superposed state. Bohm's pilot-wave theory gives an excellent account of measurement (but, because it is deterministic, it cannot give a plausible account of consciousness and agency).

According to the **many-worlds theory**, only the wavefunction (which never collapses) exists. The measuring device always exists in a superposed state. Moreover, the superposition has become ever more extreme since the Big Bang, and will continue to do so in the future. Many-worlds theory is vague about when the universe splits into superposed states, how the classical world is recovered from this superposition, and on what rational basis can the probabilities of experimental results be derived.

Measurement in GRW

According to GRW, a measuring device, such as that in the example above, cannot remain in a superposition for more than the minutest fraction of a second. The reason is that the pointer consists of a multitude of atoms. Moreover, if one particle of the pointer is at the **Start** position (say), then all of the other particles in the pointer will be, near enough, at the same position.

First, **consider the pointer in isolation**. We are only interested in the approximate position of the pointer as a whole. Is the pointer in the vicinity of the labels **Down**, **Start**, or **Up**, for example? Let **L** be the location of the pointer, and let P_1, P_2 ..., P_N be the positions of the many particles that comprise it. Then, thinking classically for the moment, $P_1 \simeq \mathbf{L}$, $P_2 \simeq \mathbf{L}$ ..., $P_N \simeq \mathbf{L}$. Thinking in more quantum terms, the only region of configuration space for which the wavefunction of the pointer is non-zero, i.e. for which $\Psi(P_1, P_2, ..., P_N) > 0$, is where $P_1 \simeq P_2 \simeq ... \simeq P_N$ ($\simeq \mathbf{L}$). Because the pointer is a solid, these atoms are constrained (entangled) in such a manner as to closely maintain their relative positions. These particles evolve according to the Schrödinger equation which, if it continued to apply for a long duration, would, to express things poetically, "dissolve the pointer into an extended, vague cloud of mere possibilities."

The rules of GRW prevent this latter possibility. Although the chance that any particular particle will localise is minuscule, there are so many (very roughly 10^{24}) particles in the pointer that one of them will localise every 10^{-7} seconds on average. Because our labelling of particles is arbitrary, we can assume that it is particle labelled P_1 that localises first. It must localise to new position $P_1 = \mathbf{M}$, where **M** is very close to **L**. Because of the spatial entanglement between the particles of the pointer, all of the other particles must also localise near **M**: $P_2 \simeq \mathbf{M}$ ..., $P_N \simeq \mathbf{M}$. A very rapid succession of such localisations, involving different particles, ensures that, to an excellent approximation, any solid macroscopic object maintains its shape, and has a well-defined trajectory. We will write the approximate location of the pointer (given by the locations of its constituent particles) as $|\mathbf{L}\rangle$; for example, the pointer might be at position $|\mathbf{Up}\rangle$ on the device. With slight modifications, essentially the same argument can be applied to semi-rigid macroscopic objects such as cats. Ghirardi (2007, pp. 410-15) presents the argument in greater and more rigorous detail.

Consider the initial and final states of the entire system, which consists of both the particle being measured and the measuring device. Call these states $\Psi_{initial}$ and Ψ_{final} respectively.

Let '→' symbolise the time evolution of the wavefunction. Suppose, first of all that the particle is spin up: $|\mathbf{s}\rangle = |\mathbf{z}\uparrow\rangle$, then by measuring device rule 1

$$\Psi_{\text{initial}} = |\mathbf{z}\uparrow\rangle\ |\mathbf{Start}\rangle \rightarrow \Psi_{\text{final}} = |\mathbf{z}\uparrow\rangle\ |\mathbf{Up}\rangle \qquad (2)$$

On the other hand, suppose that the particle is spin down: $|\mathbf{s}\rangle = |\mathbf{z}\downarrow\rangle$, then by rule 2

$$\Psi_{\text{initial}} = |\mathbf{z}\downarrow\rangle\ |\mathbf{Start}\rangle \rightarrow \Psi_{\text{final}} = |\mathbf{z}\downarrow\rangle\ |\mathbf{Down}\rangle \qquad (3)$$

So, if the particle has an arbitrary spin, $|\mathbf{s}\rangle = c\ |\mathbf{z}\uparrow\rangle + d\ |\mathbf{z}\downarrow\rangle$, then, by rule 3 the evolution of the entire system is

$$\Psi_{\text{initial}} = (c\ |\mathbf{z}\uparrow\rangle + d\ |\mathbf{z}\downarrow\rangle)\ |\mathbf{Start}\rangle \rightarrow \qquad (4)$$

$$\Psi_{\text{final}} = c\ |\mathbf{z}\uparrow\rangle\ |\mathbf{Up}\rangle + d\ |\mathbf{z}\downarrow\rangle\ |\mathbf{Down}\rangle$$

If we ignore the localisations, then the pointer has obeyed the Schrödinger equation at all times, and its final state is a superposition of $|\mathbf{Up}\rangle$ and $|\mathbf{Down}\rangle$. But in truth, according to GRW, the pointer has made many localisations. It always remains a quasi-classical object with an excellently defined trajectory. It has drifted from the **Start** position to one of the final positions, either **Up** or **Down**. Moreover, according to measuring device rule 3, the probability that it will come to rest at **Up** is $|c|^2$, and the probability that it will come to rest at **Down** is $|d|^2$.

The collapse of particle $|\mathbf{s}\rangle$, from being in a superposition $|\mathbf{s}\rangle = c\ |\mathbf{z}\uparrow\rangle + d\ |\mathbf{z}\downarrow\rangle$, to one of its basis states, **either** $|\mathbf{z}\uparrow\rangle$ **or** $|\mathbf{z}\downarrow\rangle$, occurs very early on in the measuring process: before the pointer has moved by a distance σ towards either the label **Up** or **Down** respectively. Although the pointer contains numerous particles, and the source is only one, it is the source $|\mathbf{s}\rangle$ (by way of its arguments c and d) that influences the sequence of GRW localisations as the measuring device evolves from its initial to its final state. This is surprising at first sight – one might expect the huge device simply to knock particle $|\mathbf{s}\rangle$ into its final state. But the GRW rules reveal that the single particle acts on the entire measuring device in a manner analogous to a ju-jitsu throw.

Non-measuring devices

As defined here, a measuring device must contain a macroscopic moving part. But consider Figure 4.3, *The Stern-Gerlach apparatus*. Suppose that the target is a fluorescent screen, chemically formulated in such a manner that a spin-½ particle hitting the target at a particular location will cause about 10 atoms to be excited at this location, and hence the same number of photons to be emitted from there. This apparatus contains no large parts that move through macroscopic distances, and so is not a measuring device, as this has been defined here. Let us call the region of the target labelled "z-spin up" in Figure 4.3 "**UP**" for short. Similarly, the target region labelled "z-spin down" will be called "**DOWN**".

If the source particle is in a superposition, say

$$|x\uparrow\rangle = \frac{1}{\sqrt{2}}\,|z\uparrow\rangle + \frac{1}{\sqrt{2}}\,|z\downarrow\rangle \qquad (5)$$

Then according to GRW the target screen will be in a superposition of "atoms excited in target region **UP**" and "atoms excited in region **DOWN**". Moreover, the photons emitted from the target will be in a superposition of "photons emitted from region **UP**" and "photons emitted from **DOWN**". In this particular example, the probability associated with each term in the superposition is ½.

There is a problem here that was first pointed out by David

Albert and Lev Vaidman (Ghirardi, 2007, pp. 428-33). An observer watching an experiment consisting of a repeated sequence of such trials, each performed with an |x↑⟩ particle, actually sees a spot of light at location **UP** for half of the trials, and at location **DOWN** for the remainder. Thus, when the observer looks, the superposition is found to have collapsed into one of the states: **either** "photons emitted from **UP**" **or** "photons emitted from **DOWN**". Does this mean, as Albert and Vaidman claim, that human consciousness must have caused the collapse? Ghirardi, Rimini and Weber deny this: their interpretation has been designed with the express purpose of eliminating any mention of human consciousness from the physics of the world. In the case of this experiment, Ghirardi remarks that the collapse will still occur according to GRW theory if – instead of human observers – measuring devices **as defined here** are used to detect the photons (p. 431).

But, if the experiment is performed with a human observer, the question remains: What is it about the physics of a human being that causes the collapse, and explains their experience? Ghirardi writes that this is "a real problem and causes an interesting challenge for GRW theory" (p. 431). He and others solved the problem by giving details of how neurons within the optic nerve function: Nerve cells are specialised in order to transmit information along their axons in the form of electrical signals. But they aren't merely electrical devices. Neural firing is an electrochemical process in which it is essential that many particles (sodium and potassium ions) pass from within the axon to the exterior (and back again), a distance of at least $\sigma = 10^{-5}$ cm. The time needed for this to happen is about a hundredth of a second – less than human consciousness can distinguish. In this example, the particles are not in solid form, but this is unimportant: what is crucial is that many of them must act in unison for the neuron to fire. Ghirardi gives more complete details (pp. 431-33).

Ghirardi makes it clear that he makes no attempt to explain human consciousness. He merely affirms that **the stream of human consciousness is *somehow* associated with the main bulk of the brain, downstream of the optic nerve**. If this weak and uncontroversial assumption is granted, then standard GRW theory explains why – given that we have visual experiences – these have the character they do: In each trial, particles within the optic nerve cause the superposition to collapse, so that what we see (i.e., experience) is a spot of light coming from a single specific spot on the target.

Standard GRW does not explain the more fundamental question as to how or why we have experiences. According to pan-idealist GRW, which proposes to explain consciousness, the statement in bold above remains true.

The concept of macroscopic in GRW

The concept of *macroscopic* is well-enough defined in GRW theory: it is cashed out in terms of constants σ and τ. Systems containing a large number of particles, say N, will have a well-defined trajectory, and behave more classically. For a solid body, we can even give an expression for the indeterminacy in its trajectory in terms of N. Systems with few particles will behave in a more quantum manner. The theory treats all systems, containing any number of particles, large or small, as being subject to the same laws. This is much more satisfactory than standard (Copenhagen) quantum mechanics, in which measurements that cannot be defined divide the world at arbitrary and unspecifiable locations, into two realms, which obey utterly different laws.

Identical particles, entanglement, and experients

Fermions

Fermions are fundamental particles having half-integer spin.

They are the ultimate constituents of matter. The most important of these for chemistry are the electron, which has charge = -1, the proton with charge = +1, and the neutron with zero charge. Strictly speaking, protons and electrons are fermionic because each of them is composed of three quarks, but for the purposes of chemistry, they can be treated as fermions. All of these particles have spin ½.

A proton and a neutron are nearly equal in mass; this is about 1,836 times the mass of an electron. The compact nucleus of any atom is formed out of protons and neutrons. Around this is a large cloud of electrons. If, as is often the case, the number of electrons is equal to the number of protons, then the atom is electrically neutral. But if, as can readily happen, an electron is lost or captured, then the atom is said to be *ionised*.

Identical fermions and entanglement

Suppose we have a pair of fermions, call them 1 and 2, and we swap them over, then the rules of quantum mechanics state that

$$\Psi(\mathbf{1}, \mathbf{2}) = -\Psi(\mathbf{2}, \mathbf{1}) \qquad (6)$$

Here **1** and **2** (in bold) are symbols representing the totality of variables (positional and spin) pertaining to the two particles. If these particles are physically different, say that they are a proton and an electron, then this formula is unproblematic. But if these particles are identical, say both are electrons, then these particles **must** be entangled. To see this, we suppose that they are not entangled, then by Schrödinger's definition (1935a, p. 556), Ψ can be expressed as a product of two functions, f and g.

$$\Psi_{\text{naive}}(\mathbf{1}, \mathbf{2}) = f(\mathbf{1})g(\mathbf{2}) \qquad (7)$$

On the other hand, if we were to swap particles 1 and 2, and hence all their respective arguments **1** and **2**, we obtain

$$\Psi_{naive}(\mathbf{2}, \mathbf{1}) = f(\mathbf{2})g(\mathbf{1}) \tag{8}$$

And these last two expressions are not the negative of one another. Instead, we must consider the antisymmetric state

$$\Psi(\mathbf{1}, \mathbf{2}) = \frac{1}{\sqrt{2}} [\Psi_{naive}(\mathbf{1}, \mathbf{2}) - \Psi_{naive}(\mathbf{2}, \mathbf{1})] \tag{9}$$

$$= \frac{1}{\sqrt{2}} [f(\mathbf{1})g(\mathbf{2}) - f(\mathbf{2})g(\mathbf{1})]$$

You can check for yourself that this expression satisfies (6). Because of the subtraction in (9), there is an inevitable entanglement between every particle in the entire universe that belongs to the same fermionic type. This was dramatized by Jean-Marc Lévy-Leblond, who wrote that entanglement is such that "the electrons of my body [are entangled] with those of every inhabitant of the Andromeda galaxy" (quoted in Ghirardi, 2007, p. 340).

Now let us sketch a situation with a pair of electrons where we also consider their spins. Suppose electron 1 is to be measured in region A, and its position is given by $f_A(\mathbf{1})$, where this function is zero outside of region A. Moreover, its z-spin happens to be up: $|z\uparrow\rangle_1$. Similarly, electron 2 is to be measured in region B, and its position is given by $g_B(\mathbf{2})$, where this function is zero outside of region B. Moreover, its z-spin happens to be down: $|z\downarrow\rangle_2$. Naively, we could write this situation as

$$\Psi_{naive}(\mathbf{1}, \mathbf{2}) = f_A(\mathbf{1})\ |z\uparrow\rangle_1\ g_B(\mathbf{2})\ |z\downarrow\rangle_2 \tag{10}$$

This equation can be regarded as a more explicit variant of (7). The subscript A reminds us of the region where f is non-zero; and the spin has been separated out from the arguments **1** of electron 1. Similar remarks apply to electron 2. This is not a valid wavefunction, and so we have to make it antisymmetric as before

$$\Psi(\mathbf{1}, \mathbf{2}) = \frac{1}{\sqrt{2}} f_A(\mathbf{1}) \mid z\uparrow\rangle_1 \, g_B(\mathbf{2}) \mid z\downarrow\rangle_2 - \frac{1}{\sqrt{2}} f_A(\mathbf{2}) \mid z\uparrow\rangle_2 \, g_B(\mathbf{1}) \mid z\downarrow\rangle_1 \qquad (11)$$

This is a more detailed variant of (9). We can see from the first term on the RHS of (11) that a measurement to detect the presence of an electron in region A will find the particle labelled 1 there (with probability ½); moreover, it will have z-spin up. From the second term, we can see that for the remaining half of the experimental trials, the particle we have labelled 2 will be found in region A, and it will also have z-spin up. Since we can never distinguish electrons by any experimental method, the empirical content of equation (11) is that an electron always arrives in region A, and that it is always z-spin up. A similar analysis shows that an electron always arrives in region B, and is invariably z-spin down. In any given trial we have no idea as to which electron goes to region A, and which to B.

The simpler, but not strictly valid, equation (10) describes the same physical situation: an electron can be found in each of the regions; the one within region A is spin up, and the other, at B, is spin down. The sole way in which this equation is misleading is that it wrongly suggests that the electron labelled 1 always goes to A (and that labelled 2 goes to B). The truth is that we have no physical way of labelling electrons or experimentally distinguishing them.

Here I will make a couple of definitions. Two or more fermions of the same type are *trivially entangled* if their entanglement arises solely from their indistinguishability. Trivially entangled fermions have a wavefunction that can be expressed as a product which is then made antisymmetric. This process is shown in equations (7) and (9), and a more detailed variant of it is given in (10) and (11).

Two or more fermions of the same type whose entanglement is non-trivial I will call *radically entangled*. For example, a pair of electrons that feature in an Aspect-type experiment will be radically entangled because they are simultaneously emitted from a common source. Each pair will proceed in opposite directions because of conservation of momentum; moreover, they will have opposite spins, by conservation of spin. Radically entangled particles can cause correlations between physical goings-on in spacelike separated regions A and B, as was explained towards the end of chapter 5.

Trivial entanglement does have some important effects; notably in regard to chemistry. An important consequence is that no two fermions can be in exactly the same state. For example, suppose electrons 1 and 2 were in the same state (their wavefunctions were both f), then equation (9) would reduce to

$$\Psi(\mathbf{1}, \mathbf{2}) = \frac{1}{\sqrt{2}} [f(\mathbf{1})f(\mathbf{2}) - f(\mathbf{2})f(\mathbf{1})] = 0 \qquad (12)$$

But this equation says that Ψ is identically zero everywhere. This certainly isn't a valid wavefunction because it claims that these electrons are nowhere to be found in the universe. This result, which affirms that no two fermions can exist in exactly the same physical state, is the *Pauli Exclusion Principle*. In any given atom there is a certain minimal energy, called the *ground state*, which an electron can have within this atom. By the Pauli Exclusion Principle, two electrons can exist within the ground state, because they are distinguishable by having opposite spins. A more sophisticated analysis shows how the number of electrons is built up within an atom of any given mass. Very roughly speaking, electrons fill up 'shells' of increasing energy. In a given atom, the final shell might not be filled, but if it is, then the atom is particularly unreactive. Trivial entanglement also plays a role in explaining many key features of the Periodic Table of the elements, which is paramount for all of chemistry.

Trivial entanglement is benign

Ghirardi emphasises that trivial entanglement is "absolutely benign". He goes on to affirm that, with regard to the statement of Lévy-Leblond:

> [The trivial] entanglement of electrons in my own body with those of the inhabitants of the Andromeda galaxy has a unique consequence that it is no longer legitimate to say that these are "my" electrons and those are "theirs." But, thanks to the effect of pure and simple antisymmetrization, nothing changes with regard to the objective fact that here on Earth there exists a being that coincides with me, and on Andromeda there is another one that would be very different.
> (Ghirardi, 2007, pp. 342-3)

This can also be seen in our discussions of the distribution of the electrons within an atom, or of the covalent bonding of molecules. These were perfectly complete and satisfactory without any need to mention the fact that the electrons within particular atoms and molecules here on Earth cannot be distinguished from those on Andromeda.

Bosons in brief

Bosons are fundamental particles having integer spin; they are force carriers. The most important of these for our purposes is the *photon*, or particle of light, which has spin = +1, and whose (rest) mass and charge are both zero. Suppose we have a pair of photons with arguments **1** and **2**. The wavefunction of bosons is **symmetric** when these arguments are swapped:

$$\Psi(\mathbf{1}, \mathbf{2}) = \Psi(\mathbf{2}, \mathbf{1}) \tag{13}$$

(Contrast this with the minus sign that is present in equation (6) for fermions.) Suppose these photons are unentangled, having separate wavefunctions f and g respectively. The obvious candidate wavefunction for such a pair would be

$$\Psi_{naive}(\mathbf{1}, \mathbf{2}) = f(\mathbf{1})g(\mathbf{2}) \tag{14}$$

But this is not valid for $f \neq g$ because this would not satisfy (13). We correct this by symmetrizing (14) to give

$$\Psi(\mathbf{1}, \mathbf{2}) = \frac{1}{\sqrt{2}} [\Psi_{naive}(\mathbf{1}, \mathbf{2}) + \Psi_{naive}(\mathbf{2}, \mathbf{1})] \tag{15}$$

$$= \frac{1}{\sqrt{2}} [f(\mathbf{1})g(\mathbf{2}) + f(\mathbf{2})g(\mathbf{1})]$$

Note the strong analogy between the above three equations, and equations (6), (7) & (9). The minus signs in the fermion equations have been replaced throughout with plus signs in the boson equations. Trivial versus radical entanglement can be defined for bosons in an analogous manner as for fermions. Trivial entanglement remains benign. Because of the change in sign, the experimental statistics and characteristic behaviours of fermions and bosons are different from one another. For example, bosons do not obey the Pauli Exclusion Principle because (15) is not identically zero everywhere when f = g (contrast this latter situation with (12)). A pair of photons is typically at least trivially entangled. They are only completely unentangled if they are in exactly the same state (f = g); this occurs in a laser beam, for example.

The above discussion of identical particles is based on Ghirardi (2007, chapter 14, especially pp. 341-2). Much greater detail on this topic may be found in Griffiths and Schroeter (2018, chapter 5).

Experients

According to physicalism, it makes no sense to assert that electron 1 is over here whereas electron 2 is in Andromeda: there is no physical test that will enable us to make the distinction; and – **by the physicalist hypothesis** – every real difference must be grounded in some physical distinction.

In contrast, according to pan-idealism the objective (true) physics of our universe is not fundamental. Rather, it has a deeper grounding in the percepts of experients and their volitions. This has the consequence that it is meaningful to assert that two experients may be absolutely indistinguishable physically, while still retaining their distinct identities. For example, the "something it is like to be" of experient 1 is in some primitive way an experience of being on Earth; whereas that of experient 2 is in some primitive way an experience of being in Andromeda.

As will be explained below, every radically entangled system is a physical manifestation of an experient.

Introducing the pan-idealist GRW universe

One of the intentions of standard GRW theory, is to eliminate any role for observers within physics, so as to provide a purely objective account of the universe. In contrast, the aim here is to understand consciousness; in particular, the active role of mental agency in the functioning of the cosmos. Our task therefore is to fully integrate ourselves – in a naturalistic manner – into the world. By "in a naturalistic manner" I mean that there must be nothing so exceptional about human beings that we belong to a different category of existence from anything else in the universe. (*Substance monism* is the desirable philosophical principle that everything that exists has the same general character. Pan-idealism is an example of substance monism.) This motivates what follows.

In standard GRW's theory, particles are the passive targets of causeless measurement-like localisation events, or flashes. But GRW can readily be modified to *pan-idealist GRW*, in which *agents* actively **cause** the flashes. "Agents" is just a synonym for experients: the only reason for the alternative name is that this chapter focuses on the active choices of the experients, rather than on their percepts, which are passive in character. The evolution of the pan-idealist GRW universe is explained in detail over the next three figures.

The wavefunction Ψ of the world is denoted, writing its arguments (its inputs) explicitly, as $\Psi(\mathbf{q}_1, \mathbf{q}_2, \ldots, \mathbf{q}_N; t)$. The lowercase t expresses how the wavefunction would evolve according to the Schrödinger equation if no localisations occur. With standard GRW – and likewise with pan-idealist GRW – there are flashes which cause the wavefunction Ψ of the world to change ('collapse'), becoming another. In the account which follows, the superscript uppercase T, as in Ψ^T, is used to keep track of the change from Ψ to a new Ψ, caused by flashes. It is important to understand the distinction between t and T. Lowercase-t is used to describe how a single wavefunction evolves according to the Schrödinger equation in the absence of flashes. Uppercase-T, on the other hand, keeps track of how one wavefunction changes into another, because of flashes.

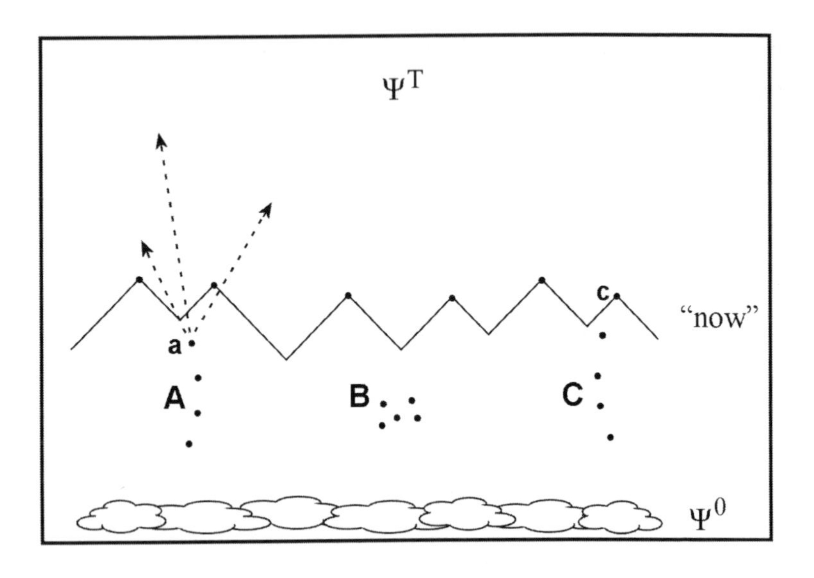

Figure 10.2: Cosmic time T in a pan-idealist GRW universe

Figure 10.2 shows the current (at cosmic time T) state of a pan-idealist GRW universe. Consider the following definitions: *flashes* are GRW localisation events **that have already occurred** by time T. Function Ψ^0 (symbolised by 'clouds') is the initial wavefunction for the universe, before any flashes have happened. The *cosmic past* is defined by combining together (in mathematical jargon 'taking the union of') the past light cones of all the flashes. The past is unalterable and is given uniquely by Ψ^0 and all of the flashes. The *cosmic present*, or *'now'*, is the upper boundary of the cosmic past (and, strictly speaking, it forms part of the latter). 'Now' is shown by the zigzag line segments, which are past-directed light cones. All flashes that lie in the cosmic present are spacelike separated. The *future* does not yet exist, but is what may potentially happen – consistent with objective physics – given a complete account of the current state of the universe. There are many different possibilities as to what the future may turn out to be. It lies above the cosmic present in Figures 10.2-4.

Ψ^T is the wavefunction of the universe at cosmic time T. It is fully determined by the universe's initial state Ψ^0, and its history in terms of all the flashes (up until cosmic time T). Ψ^T fully encodes:

- All possibilities for the universe's future
- The universe's current state of radical entanglement

As was explained earlier, Ψ^T can be expressed uniquely as a suitably symmetrized or antisymmetrized product:

$$\Psi^T = F[\, \alpha_1\, \alpha_2\, \alpha_3 \,\dots\,] \approx \alpha_1\, \alpha_2\, \alpha_3 \,\dots \qquad (16)$$

Strictly speaking, each system of particles α should have a superscript T, but this is omitted here. Here F is an operator that suitably symmetrizes or antisymmetrizes the individual α's so as to give the valid wavefunction Ψ^T. If we disregard F then we are left with a naive product that ignores trivial entanglement. With care it is often safe to do this because trivial entanglement is benign. F is an all-purpose recipe: "make this naive product into a valid wavefunction". F only becomes a specific – albeit highly complicated – operator once the exact particle content of each of the α's, and the specifics of their radical entanglement, have been given. The product in (16) has the following properties:

i. Every particle in the universe is in exactly one of the α's.
ii. No pair of distinct α's are radically entangled with one another.
iii. Each α describes either: the state of a system of radically entangled particles; or the state of a free (unentangled) particle. In other words, no α can be factorised into a product satisfying (ii).

Suppose, for simplicity, the universe contains a fixed and finite number of particles 1, 2, 3 ..., N. As explained in the section on experients, within pan-idealism it is valid to assert that a specific particle belongs to a particular entanglement. At this point I am going to introduce some notation of my own by writing (say):

$$\alpha_1 \approxeq \{1, 4, 7\} \qquad (17)$$

Here the brackets { } are to be understood to mean that "α_1 is some radical entanglement of particles 1, 4 and 7." The RHS does not specify the manner in which these particles are entangled, whereas this information would be given explicitly in the mathematical expression for the wavefunction α_1: If **1**, **4** and **7** are the arguments pertaining to particles 1, 4 and 7, then we could (if we wish) write α_1 as $\alpha_1(\mathbf{1}, \mathbf{4}, \mathbf{7}; t)$. The richer information content on the LHS is why the symbol '≊' rather than '=' is used in (17). Property (iii) above asserts that α_1 cannot be expressed as a product such as {4} {1, 7}.

Using this notation, the wavefunction of equation (16) might be for instance:

$$\Psi^T \approx \alpha_1\, \alpha_2\, \alpha_3 \ldots \approxeq \{1, 4, 7\}\, \{2\}\, \{3, 6\} \ldots \qquad (18)$$

Here the symbol ≈ can be read as "is roughly"; and ≊ can be read as "is even more roughly". In this equation, $\alpha_1 \approxeq \{1, 4, 7\}$ has already been discussed. Next, $\alpha_2 \approxeq \{2\}$, so particle 2 is not radically entangled with any other: it is said to be *free*. Finally, $\alpha_3 \approxeq \{3, 6\}$ is a radically entangled pair of particles. Each of the entanglements of expression (18) contains only a small number of particles. Such small entanglements are frequently found in our universe, but entanglements can also contain vast numbers of particles.

In Figure 10.2, timelike separated flashes – like footsteps in the snow – indicate the paths of particles, as seen at A and C. I am going to call both the particle and the track it makes by the same name. So we have particle A whose track is A. The same applies to C.

Track C ends at flash c in the present, particle C is therefore currently well-localised. Track A does not end in a flash at the present; particle A's last known position was at a (the most recent flash on the track). This point could in principle be anywhere in the near or far distant past (it might have occurred a few seconds ago, or just after the Big Bang). Particle A's current location does not exist, but information about possible locations in the future are encoded in Ψ^T, as symbolised by the dotted arrows. Sometimes, as shown near B, there might be a cluster of flashes that cannot be separated into obvious tracks.

Agents

According to standard GRW, particles usually follow the Schrödinger equation, but are occasionally, at random times, subject to localisations or "flashes", which move them close to particular spatial locations. In all of this the particles are entirely passive. How are we to amend this to a dynamic in which agents, acting out of free will, cause flashes that change the course of history of the universe? The problem lies first in characterising the agents appropriately; and second in specifying a means of ensuring that the choices of the agents are mutually consistent.

Key idea: We are going to *recognise* **every** radically entangled system of particles (also every free particle) discussed in the previous section (i.e., each of the α's at cosmic time T) as being the physical manifestation of an *agent*.

I say "recognise" rather than "identify" because we are not taking the agents to be identical to (in the formal sense of 'to be one and the same as') anything physical or formal. Agents are mind-like, and they are fundamental; whereas entanglements are physical, and hence secondary. To give an analogy: Solely on the basis of physical evidence I recognise my friend Jim to be a person (i.e., he is, in his fundamental and intrinsic nature, a mind) who enjoys experiencing extremely hot curries. But Jim's mind is not one and the same as any physical structure. As explained in chapter 8, this does not imply that I am a dualist: According to pan-idealism all physical facts can be reduced without remainder to experiential facts. I can recognise Jim as being physically manifest in a particular, very complicated system of particles. The distinction between physical systems and agents is crucial. Despite this I am going to call each radically entangled physical system, and the corresponding experiential agent that underlies it, by the same name, α. This is analogous to me pointing to a particular system of particles and saying, "That's Jim!"

The next step is to specify the dynamics of the pan-idealist GRW universe in such a manner as to ensure that the free choices of the agents are mutually consistent.

The dynamics of the pan-idealist GRW universe

Each agent, ignoring the others, makes a choice as to where and when in the future it wishes each of its constituent particles to land. Typical agent α_j has enough information to ensure that this (combined) choice is consistent with:

1. The current state of the universe, Ψ^T
2. The precise nature of the entanglement of the particles within the agent α_j

This entanglement is given by the full and explicit mathematical expression for the wavefunction α_j. For example: $\alpha_1 = \alpha_1(\mathbf{1}, \mathbf{4}, \mathbf{7}; t)$, where, on the RHS, label α_1 is assumed to have been written out as an explicit mathematical formula.

3. The GRW rules

These rules – recalled at the beginning of this chapter – **do not determine** what choices each agent will make; they merely **describe** statistical propensities as to what choices each agent will make. (This is the real meaning of the Born Rule.)

4. Attempted choices are indeed all in the universe's future at cosmic time T

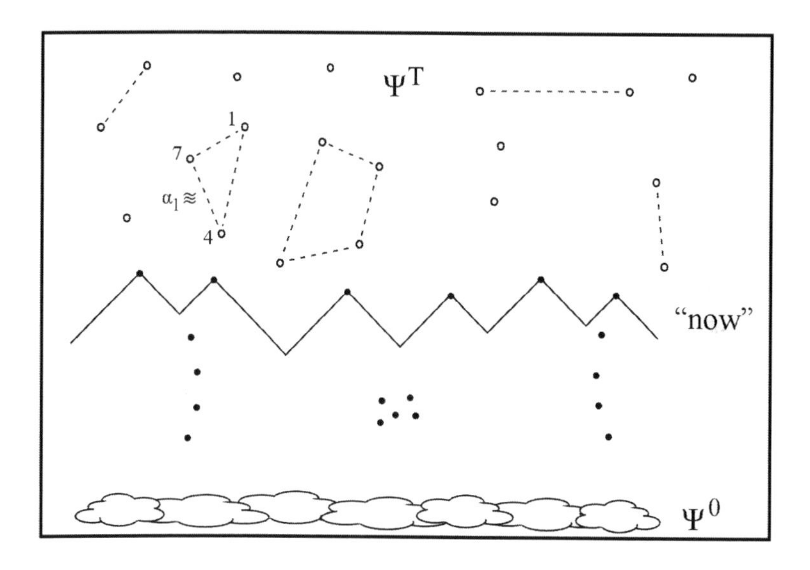

Figure 10.3: At cosmic time T, independently of one another, every agent makes an attempt to land all of its constituent particles at specific locations – of its own choosing – in the future. The agents shown here are: six free particles; three entangled pairs; an entangled trio (α_1); and an entangled quartet.

Figure 10.3 shows agent α_1's combined choice of where it wants its three particles to localise themselves ("flash") in the future. Every agent will invariably have an infinite number of Alternative Possibilities for its combined choice, consistent with the four rules above.

The figure depicts where all of the agents have chosen to **attempt** to flash (**o**). Shown here are six free particles, three entangled pairs, an entangled trio (our α_1), and an entangled quartet. These attempted choices will almost never be consistent because each agent has made its choice independent of what any of the other agents are doing. For example, suppose that an agent, say β, has chosen to attempt to flash one of its particles within the future light cone of another agent's attempted flash. The latter flash will have already disturbed Ψ^T, thus invalidating agent β's choice according to the four rules. How are such conflicting choices to be arbitrated?

Reconciliation is done at the level of particles rather than agents, and is illustrated in Figure 10.4. Any attempted flash that appears in the future light cone of any other attempt is deemed to have failed, for the reason given in the previous paragraph. Such flashes are eliminated (×). The remaining flashes are successful. They are joined together by their past light cones to produce a new "now". (The previous cosmic present moment I've re-named "just now".)

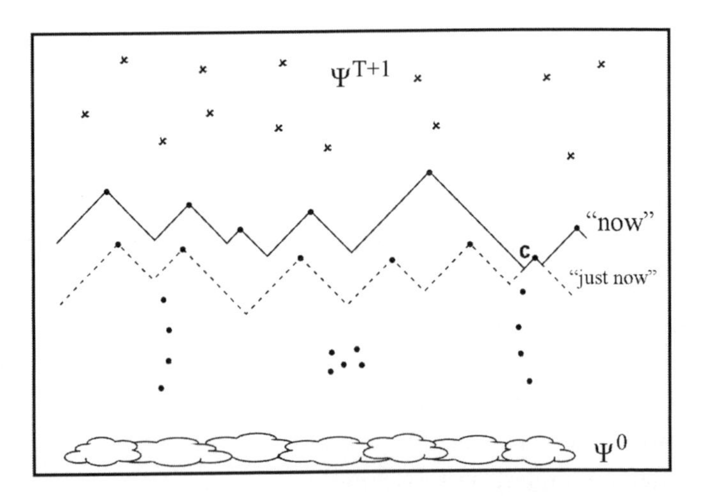

Figure 10.4: Resolving inconsistencies. Attempted flashes in the future light cone of others are deleted (×). The remainder are accepted, producing a new "now". We arrive at cosmic time T+1.

To the right of Figure 10.4 there is a flash, **c**, that is common to both "now" and "just now". (It was given this label in Figure 10.1.) This flash is associated with a particle, call it P. This situation is perfectly okay. It arises when there have been no successful flashes by any particle (P included) in the future light cone of flash **c**. (It is irrelevant whether or not P is radically entangled with other particles.)

Comparing Figures 10.2 and 10.4 we see (ignoring "just now" and the deleted unsuccessful flashes) that we have an analogous situation, but one in which T has been replaced by T+1. Because of the successful flashes, Ψ^{T+1} is slightly different from Ψ^{T}. This has the implication that the set of agents at time T+1 are a little different from those at time T. A system of particles α can become radically entangled with one or more others. Likewise, α might split into separate pieces that are only trivially entangled with one another. The same applies to the underlying agents.

Cosmic Time

The parameter T, which labels how the wave equation changes as a result of a new layer of successful localisations, has the character of a universe-wide, cosmic time. It is naturally rather than arbitrarily defined. Moreover, this definition is unambiguous and precise. But it is one which we can never have the knowledge to calculate. We do not and cannot have exact knowledge of Ψ^{T}, let alone full and explicit mathematical expression for the wavefunction of every other agent α_j.

For example, in a trial within an Aspect-type experiment, the measurement event at A might occur at an earlier cosmic time than the one at B, or vice versa. These events might even occur at the same cosmic time. According to pan-idealism, there is a definite fact as to which of these three possibilities occurs, even though we have no way to distinguish them, and they have no effect on the trial outcome. Another example is that there is a well-defined mathematical criterion for a flash to be in the past, present or future at cosmic time T, even though there is no possibility for humans to calculate it. Nonetheless it is true that, by rule 4, all attempted flashes are, as a matter of fact, in the future.

In my view, the fact that the concept of cosmic time is self-consistent and well-defined is sufficient for its physical truth; it is irrelevant whether or not anyone could ever calculate it – I am not an empiricist.

How far separated are these time slices T? One might think that in the presence of macroscopic objects, such as those that are found in abundance on Earth, the slices are extremely close together. But the example of the S-G apparatus with fluorescent screen should give us pause. Moreover, suppose we develop a clock that can measure extremely small time intervals; might it be the case that this device itself creates these small intervals? The question remains hard to answer. Pan-idealism and GRW do not eliminate all of the weirdness of quantum mechanics. This is not in itself a problem; it merely reflects the way our world happens to be.

Process

According to pan-idealism, there is a *process* by which the universe as a whole gradually comes into existence over cosmic time. The past and the present physically exist, but the future consists only of potentialities – of possibilities that may or may not come into being. As cosmic time passes, then there is ever more of the universe 'to date' that concretely exists. It comes into being in a way analogous to a sedimentary rock, to which layers are gradually added.

Alfred North Whitehead (1929) is the best-known proponent of a – somewhat different – process cosmology. In contrast, Einstein's general relativity is a "block universe" in which past, present and future are indistinguishable in their manner of existence (1920).

Cosmic time and general relativity

Bassi and Ghirardi (2007, p. 177) explain an Aspect-type experiment in terms of a relativistic GRW model. They describe the instantaneous collapse of the wavefunction and say (emphasis added), "Now, let us suppose that the position of particle a is measured **slightly before** that of particle b." Unfortunately, the underlined phrase is not strictly meaningful in a relativistic context. Pan-idealist GRW has the advantage that it gives precise (albeit untestable) meaning to this statement in terms of the naturalistic cosmic time T.

Looking back over this section, it is even possible to describe this cosmic time T in terms of **standard** GRW theory, provided the passive voice is used throughout: "Associated with each radically entangled subsystem α_j, and ignoring all others, there are potential flashes of each of its particles, consistent with rules 1 through 4" and so on. Describing cosmic time in this manner should be of interest, even to those physicists who entirely reject my theory of mind.

The combination problem

The *combination problem* asks: How do physical ultimates having primitive mental properties combine to form unified minds of the sort that we know ourselves to be? This is widely regarded as the most intractable problem in panpsychism (and pan-idealism), and the anthology edited by Brüntrup and Jaskolla (2017) devotes five out of sixteen chapters to it. Some academics claim that no answer is possible, and have concluded that panpsychism should therefore be abandoned as untenable. Previous attempts by panpsychists to solve the combination problem have been contrived, and have also been vague in terms of their relation with physics. Most proposals do not engage with the specifics of any science.

The pan-idealist GRW universe gives a fully naturalistic account of the combination (and dissolution) of agents, without recourse to special laws outside the domain of physics. Agents attempt to act in the world at cosmic time T. Some fail in this attempt, and some succeed, perhaps partially. As a result of their attempts, there is a slightly different state of the world Ψ^{T+1}, with a somewhat different set of agents. Some agents have split (become disentangled); some agents have combined (become entangled); and some are unchanged in terms of the particles that comprise them.

Consider a human brain in these terms: it consists of a large number of agents, most of which will be composed of a vast number of entangled particles. At time T+1, these agents will be somewhat different from those at time T. But, in principle, an all-knowing, all-seeing observer of our universe would be able to identify a particular agent B at time T+1 as being the successor of agent A at time T, solely on the grounds of the overwhelming proportion of particles that A and B share in common. In a single human brain, there would be a whole family of agents A_1, A_2, ... at time T that evolved into a corresponding family of related agents B_1, B_2, ... at time T+1. The great majority of such families would constitute extremely primitive streams of experience of which we are unaware.

That these short-lived agents/experients at T+1 are determined by the choices of the corresponding agents at time T allows them to be rightly regarded as constituting a single human person existing over time. (The analogy here is with an animal that is correctly regarded as continuously existing over the course of its lifetime, even though its constituent atoms have been replaced repeatedly with others from its environment.) The arguments about free will of chapter 9 (based on the ideas of Robert Kane) still go through.

In line with the findings of neuroscience, there may be several streams of highly developed consciousness within a single human brain, one of which may be dominant in determining behaviour on some occasions, and others dominant on different occasions. For example, I may be singing a song, which corresponds to one momentarily-dominant stream of consciousness; an aggressive dog approaches so I become alert to it, which corresponds to a second stream; I take action, perhaps running away or alternatively trying to scare the dog, and these options correspond to another two among many other possible streams. This is in essence Daniel Dennett's (1991) "fame in the brain" theory of consciousness, but cast in qualitative, idealist, and libertarian terms; rather than according to his preferred mechanistic, physicalist, and compatibilist position.

I am assuming that, in the pan-idealist GRW universe, faculties such as human memory, and cognitive abilities can be modelled in a manner essentially similar to the (broadly functionalist) accounts offered by physicalists. In particular, an account of free will – as this was defined in chapter 9 – can be given in terms of the Alternative Possibilities acting in a specific ('appropriate') way in humans possessing significant cognitive powers, good memory, and an ability to predict the likely outcomes of their actions.

The credibility of such agents

The judgement made in this book – that entangled systems (and also free particles) are the physical manifestations of agents – is in line with our intuition that an agent is something that displays evidence of an irreducible unity in its behaviour. Moreover, it should have the ability to exhibit differing behaviours when placed in physically indistinguishable environments. An entangled system of particles satisfies both these criteria *par excellence*. Furthermore, as discussed, such agents can in principle model human consciousness.

A point having considerable weight is that there is no immediate knock-down argument that would prevent the extremely primitive experients of the very early universe evolving by natural selection into creatures possessing human minds and agency. This is in contrast with the implausibility of physicalist accounts, which either deny outright that mind exists, or redefine mind in terms of specific behaviours, or argue vaguely that mind "emerges."

Philosophy contrasted with physics

Some physicists (notably Stephen Hawking) claim that philosophy is redundant because physics has taken over its role. I will argue that this is not so. First a question: Might the agents of the pan-idealist GRW universe have been defined in a different manner?

One possibility is to have a single agent, call it **Nature**, that alone (at each cosmic time T) makes all of the choices for every particle in the universe, and moreover does this consistently. There is no contradiction in the existence of such a world, but it has the serious problem that the experiences of any other entity (including humans) could not play a causal role. It would seem more consistent here to affirm that **Nature** is the sole experient; and to deny the existence of all other (irrelevant) experients. Despite this, a similar position, called *cosmopsychism*, is a popular variant of panpsychism that is under wide and active discussion (see Seager, 2020).

Standard GRW, pan-idealist GRW, and **Nature** GRW are identical in terms of their physics: they share a common spatiotemporal mathematical model. However, they do differ in other respects:

- In standard GRW, flashes occur at random, i.e. without any cause: there are no experients (whether possessing agency or no)
- In pan-idealist GRW, flashes are jointly caused by multiple experients possessing agency
- In **Nature** GRW, flashes are caused by a single experient, **Nature**, possessing agency

Philosophy discusses any conceptual distinction, including those that are physical. Because these systems are conceptually (metaphysically) different, but are identical as physical systems, this demonstrates that philosophy is broader in its scope than physics.

The models differ in terms of causes, experients, and agency, but these are philosophical rather than physical differences. Although we often speak of physical causes, in this context we mean no more than either:

1. A constant conjunction of events: For example, a brick hitting a window with sufficient velocity will always cause it to break. But, as David Hume pointed out, a constant conjunction of events does not get us to the heart of causation – it merely says that if this happens, then that will happen.
2. Or some physical law (more accurately, some mathematical regularity) is found to hold: For example, Newton's laws, such as F = ma, enable us to predict a projectile's trajectory. But systems of mathematical equations, which are of themselves purely abstract, cannot amount to physical causes.

Cause is properly linked with the philosophical concept of agency, as when my decision to lift a cup causes it to be raised. John Searle puts it this way

> The form of causation that we are discussing here... is not a matter of regularities or covering laws or constant conjunctions. In fact I think it's much closer to our commonsense notion of causation, where we just mean that **something *makes* something else happen**...
> (1984, pp. 64-5, all emphasis added)

Some academics argue that philosophical distinctions are meaningless because they cannot be tested empirically (i.e., by experiment). In contrast, I would argue that the three systems differ in the adequacy of their solutions to the mind-body problem. Standard GRW must offer a physicalist solution but, as Ghirardi admits (2007, p. 428), this is far from being attained. I claim advantages for pan-idealist GRW, but you must be the judge. In **Nature** GRW, it seems to me that the assertion that the agent **Nature** exists is irrelevant to solving the mind-body problem. Its difficulties are those of physicalism.

Assessment of pan-idealist GRW

You will want to make your own assessment. Here is a summary of my views.

Advantages

Pan-idealist GRW has several points in its favour:

- *The theory recognises that every radically entangled system of particles is the manifestation of an agent*

This is a plausible conjecture because each of these things, physical and experiential respectively, is holistic in character. (Recall from Part I that there are strong, independent arguments in favour of panpsychism.)

- *The combination problem is fully resolved*

The solution presented here is complete, precise, and plausible. Crucially, it far surpasses anything that has previously been proposed.

- *Acts of volition are precisely defined and are unexceptionable*

Successful localisation events were defined in the discussion of Figures 10.2-4. They are clearly caused (made to happen) by the agents. Contrast the precision and effortlessness of this explanation with the difficulties that all physicalists have in accounting for mental causation. A sign of the latter is that well over 90% of papers written by physicalists on the mind-body problem make no allusion to mental causation – instead they discuss perception alone.

- *The two-stage temporal process is the obvious – perhaps only – way to resolve conflicts between the choices of agents*

Such conflicts are inevitable. The process is analogous to two children wanting a biscuit. Whoever grasps it first is the one to eat it.

- *There cannot possibly be an explanatory gap*

This is because the theory is idealist – everything physical has a complete explanation in entirely mentalistic terms. Contrast this for example with Strawson's pure panpsychism, which was discussed in chapter 8.

- *The theory is unequivocally realistic*

Despite being a form of idealism, pan-idealist GRW gives a precise definition of what it is to be concretely real: it is to be an experient/agent. This is in contrast to the vanishingly thin, abstract characterisation of concrete reality (in the absence of observers) provided by physicalists. Chapter 8 discusses this.

- *The theory has specific links with contemporary physics*

These links are more robust than in any previous attempted solution to the mind-body problem. In most academic papers such links are completely absent. There is an under-examined

– or even unconscious – bias in physicalist explanations. The preconception is that explicit links do not need to be provided; they must exist in some form or another, because physicalism must be true.

Neutral facts

There are some facts about pan-idealist GRW that are neutral:

- *Many truths of pan-idealist GRW cannot be tested empirically*

For example, the theory defines an ever-increasing cosmic time T, but no experiment could ever detect it. One might take the view that every truth about the cosmos must be testable. Ironically, this view is untestable. More important considerations are that the definition of T is a natural one, and it is fully consistent with general relativity.

- *GRW theory (standard or pan-idealist) makes slightly different predictions from quantum mechanics*

These minuscule discrepancies have not been detected as yet. This is not a serious difficulty – it is in the nature of all scientific theories to be falsifiable.

Limitations

Pan-idealist GRW has some limitations:

- *It gives no description of the agents' percepts*

What we do know about is each given agent's propensities to act in a given situation – and no more. We do not even know the structural information contained within a given agent's percept. On the other hand, no mind-body theory has so far given a satisfactory answer to this question. In more standard approaches, this unsolved difficulty is related to the search for *Neural Correlates of Consciousness* (NCCs) (Chalmers, 1996b).

- *How qualia affect propensities to act remains unexplained*

Pan-idealism does have the advantage over other theories that the bare existence of qualia within an experient's percept is unexceptionable. This is because qualia are fundamental. But this is not sufficient to explain how (as they must) qualia influence my actions: I take an umbrella when I see that the sky is grey, but not when I see it is blue, and so on. Because I cannot deny the reality of such influences, qualia must be represented somewhere within the mathematical structure of the physics. In GRW, Ψ is the **only** such mathematical structure.

My provisional suggestion, hardly amounting to an idea, is extremely crude: Ψ is a function which associates a complex number with each possible configuration of particles. The structure of one's percept gives information about the configuration of entities that surround one. So, to this extent, Ψ is appropriate to represent the structure of a percept. The associated complex numbers might be an abstract representation of concrete relationships between qualia, remotely akin to a colour wheel.

- *GRW might turn out to be unequivocally false*

There are limitations to the original GRW theory: the parameters σ and τ are slightly arbitrary; there are difficulties (skated over here) in dealing properly with identical particles; and in how one is to combine GRW with general relativity. Ghirardi himself was modest in his claims for his theory. He was particularly concerned about the last problem (Ghirardi, 2007, p. 453). In contrast, Bell stated that

> The [GRW] model is as Lorentz invariant as it could be in the non-relativistic version. It takes away the ground of my fear that any exact formulation of quantum mechanics must conflict with fundamental Lorentz invariance.
> (Bell, 1987, p. 14)

Pan-idealists are not committed to the GRW model as the last word in physics. Other, related theories could be adapted, for example the continuous spontaneous localisation (CSL) theory of Pearle (mentioned in Ghirardi, 2007, p. 405). Another possibility is well worth considering...

Pan-idealism and Penrose's OR theory

In *The Emperor's New Mind* (ENM, 1989), and *Shadows of the Mind* (Shadows, 1995), Roger Penrose gives semi-popular accounts of his still-maturing *Objective Reduction* (OR) theory of quantum physics. Although, as the titles suggest, he discusses the mind-body problem extensively, he makes no pretence to have solved, or even touched upon, Chalmers' Hard Problem. Indeed, he asserts:

> I believe that the problem of quantum measurement should be faced and solved well before we can expect to make any real headway with the issue of consciousness in terms of physical action – and that the measurement problem must be solved in entirely physical terms.
> (Shadows, p. 331)

In this paragraph I consider OR theory as a purely physical theory, disregarding all of Penrose's discussion of consciousness. Penrose is a realist about the world and its contents, for example, he takes the wavefunction Ψ to be an enigmatic sign of some reality (Shadows, ch. 6). He also takes the collapse of the wavefunction to be a real physical event, occurring frequently throughout the universe, independent of any consciousness (p. 335). In each book, he gives somewhat different criteria for the collapses of superpositions, but both are based on the mass of the system under consideration (ENM, pp. 475-81; Shadows, pp. 339-46). According to the latter work, a neutron or proton can remain in superposition for 10 million years; a speck of water 10^{-3} cm in diameter would collapse from superposition in less than a millionth of a second (Shadows, pp. 340-411). OR theory is stochastic and obeys the Born Rule.

Penrose construes consciousness narrowly – as something that requires substantial cognitive abilities. In collaboration with Stuart Hameroff, he has published papers (one is reprinted in Shear, 1997) arguing that such sophisticated consciousness requires an appropriately organised ("Orchestrated") OR occurring within the brains of humans and some other creatures. They propose a partial account of the kind of physical processes needed to achieve this. Both are open to the possibility that consciousness in a broader sense (i.e., as understood in this book; sometimes called "experience" or "proto-consciousness") might be widespread or even omnipresent (panpsychism).

It should be clear that OR, regarded as a purely physical theory, can be readily adapted to pan-idealism. This is because, in its relevant characteristics, it is similar to GRW. The gravitational criteria given by Penrose have the advantage that they are less arbitrary than GRW's parameters σ and τ. Moreover, Penrose and Hameroff's position about cognition may also be carried over.

Penrose rejects any non-physicalist approach to mind, which he characterises as the position that "Awareness cannot be explained in any physical, computational, or any other scientific terms" (Shadows, p. 12). His grounds for doing this are concerns that such an approach will inevitably lead to mysticism, rejection of science, and either to irrealism about the world, or to mind-body dualism. I hope that the discussions of

Part III have shown that these worries are groundless: if, when completed, OR theory happens to be the true physics of the universe, then pan-idealists can accept this fully.

Chapter 11

Concluding reflections

Now that pan-idealism is tolerably well developed, it is presented here as an example of a metaphysical system. I want to defend traditional metaphysics, at least to the extent that it should neither be entirely superseded by analytical philosophy, nor by metaphysics conducted in an analytic manner. (Analytical philosophy has great value, but it should not be the only tool in the kit.)

Metaphors can be misleading but, now that pan-idealism has been specified without them, they are harmless: at this point you are in a position to judge for yourself to what extent they are appropriate. A couple of examples are given here. Also, because the concept of maximal intersubjective consilience is so central to pan-idealism, an example is sketched of how this is derived in a simplified, toy universe. An answer to David Chalmers' implicit critique of such consilience is given.

Before closing, there is further discussion of physicalism; and also a new topic: Pan-idealism's assessment of the extent to which, either now or in the future, digital or quantum computers might be conscious.

Pan-idealism as a traditional metaphysical system

The modern way of working in metaphysics remains analytic in tenor. The approach is to take a class of metaphysical systems that fall under some description, say **emergence**. You then compare members of that class, describing where one variant is better than another. Philip Clayton's (2004) book exemplifies this. He is an emergentist by conviction, and he discusses various ways this could be so: strong versus weak emergence; five different meanings of 'emergence'; three levels of emergence of mind;

and also four metaphysical responses to the emergence of mind.

In contrast, my aim has been to develop and pin my colours to the mast of a single, specific metaphysical system, pan-idealism, and to defend it as being (so far as I can ascertain) the truth about the world in which "we live and move and have our being." This has been the traditional way of undertaking metaphysics, as given by the examples of Descartes, Leibniz, Berkeley, and many others. A few continue in this tradition even in the present era. Notable is Timothy Sprigge, as the title of his 1983 work, *The Vindication of Absolute Idealism*, makes abundantly clear.

Why my confidence?

Why do I have such confidence in pan-idealism? After all, at most one traditional metaphysical system can be true: the remainder are necessarily false. This is the case, even for the systems espoused by the eminent philosophers just named. My answer is in several parts:

- The universe has an objective metaphysics

Although this is not provable, I am convinced it is true. Denying this assertion would be to admit the possibility that our universe is utterly absurd. But the cosmos is far too exquisitely precise and beautiful for this to be at all plausible.

- Suppose pan-idealism is plain wrong

My position is that, if the universe has an objective metaphysics, it would be craven not to seek it. If one's system is wrong, there is no disgrace in failure. This, surely, has been the attitude of all traditional metaphysicians.

- There are reasons to be confident

Pan-idealism leverages the fact that our sole knowledge of the universe comes to us through our senses and through our wilful actions: there is no other way. This applies equally to all scientific knowledge: Scientific theories and laws, whether or not they are correct, arise in human minds on the basis of experience. This has been true throughout the entire history of science. Moreover, it is impossible that science could ever change, in a manner so radical that it would avoid this undeniable fact of human existence.

I've tried, so far as I am able, to marry pan-idealism to current scientific knowledge, specifically to GRW. But what will happen if scientific theories change in the future, and GRW is found to be false? My belief, based on the argument of the previous paragraph, is that pan-idealism will be readily adaptable to novel theories that come along. As mentioned previously, Penrose and Hameroff's OR theory is one promising alternative.

- Pan-idealism has made solid progress in metaphysics

The achievements of pan-idealism were discussed at the end of the previous chapter. Another great benefit has been an improved understanding of *concrete existence*. Chapter 7 made the case that everything that is concretely real **must** be characterised by being mind-like in its intrinsic nature. It is impossible, even on the deepest reflection, to form any meaningful notion of an entity – in this or in any other universe – that is concretely real, but which totally lacks mind. This applies to all possible worlds: to exist concretely (as an individual entity) is precisely to be a centre of experience.

Traditional metaphysics

Is traditional metaphysics dogmatic?

Contrary to what one might expect, adherence to a particular system of traditional metaphysics does not make one unable to change. I will give my own experience as an example.

Physicalism: Before I became interested in the mind-body problem, I was (by unthinking default) a physicalist. I knew that there were unresolved issues in the vicinity of consciousness, but I believed that these would eventually have a scientific solution. Then I read *Consciousness Explained* (1991) by Daniel Dennett. Its wholesale rejection of phenomenology (i.e., of human experiences, such as a feeling of pain), and its replacement by heterophenomenology (the study of the verbal reports of subjects) was, in my view, the effective denial of mind. But the book made me aware of both the importance, and the extreme difficulties of the mind-body problem. It also gave the hint that this problem might be beyond the scope of science. (Heterophenomenology avoided the issue, and any other solely 'objective' approach must inevitably do the same.)

Panpsychism: The first metaphysical position that I adopted consciously was when I became a (still physicalist) panpsychist. This was not at a whim – it solved at least two problems: the intrinsic nature problem and the boundary problem. First, any entity that concretely exists, because it has a mentalistic intrinsic nature, is now distinct from a mathematical abstraction. Second, it removes the boundary problem of standard physicalism, where one is under an impossible obligation to give a principled account – in physical terms – of the boundary between insentient and experiential matter. An account of this boundary is required both in evolutionary biology, and in the development of each individual creature.

Idealist panpsychism: A second change was when I became an idealist panpsychist. I saw that the objective physics of the world (whatsoever it actually is) could – at least in principle – be defined in terms of correlating the experiences of all experients, of which human scientists are presumably a representative sample (see Ells, 2011, pp. 83-85). At that time, I made no attempt to do this explicitly – which is done for the first time in this book, in Part III.

Pan-idealism: Initially I thought that idealist panpsychism was no more than an unusual variant of panpsychism, but this does not put things in the correct proportion. Physicalist panpsychism (PP) asserts that a certain property (mind) is distributed over the entirety of a universe which is, in its essence, physical. PP is a minor change from standard physicalism (which asserts that mind is restricted to limited regions of spacetime): the disagreement is merely about how widespread a certain property (mind) is. Idealism, on the other hand, is a major metaphysical shift. It asserts that the ground of all being is mind-like and not physical in its essential character.

Agency: I have always held that we indeed possess genuine (libertarian) free will. The most compelling argument in favour of this conclusion is that, if mind had no causal powers, this would call the very existence of mind into question (epiphenomenalism was rejected in chapters 1 & 8 here). Robert Kane is a physicalist libertarian, and his 1998 work is the strongest possible defence of this minority standpoint. Despite this, in my view his position is not fully consistent with physicalism. Rather than give up on free will, it is much better to renounce physicalism – which is already highly dubious for many reasons. As shown in chapter 9, pan-idealism is fully consistent with a Kane type libertarian position. (Most philosophers reject this latter possibility, but their arguments are unsound because they are based on the presumption of physicalism.)

To sum up, it is possible to be committed to a specific metaphysical position, but to develop it on firm rational (not empirical) grounds as one learns more. It may even become necessary, based on these grounds, to take up an entirely different position.

Critiques of traditional metaphysical systems

All metaphysical systems are based on postulates that may be called into question. How does the traditional metaphysician respond to such objections? And how does this contrast with an analytic philosopher's typical response?

For example, in Seager (2020, p. 368) David Chalmers raises a potential objection to a particular form of idealism. The concern holds that "mental properties require non-mental properties behind them to causally sustain their structure and dynamics." This assertion is, of course, a flat denial of idealism.

Suppose that both the idealist and their opponent are analytical philosophers. Such an idealist's typical response to the objector would be to find additional arguments in favour of idealism, while the objector will give additional arguments against. Since the postulate of idealism – particularly when considered in isolation – is indeed very much open to question, this debate can proceed indefinitely, becoming ever more convoluted, and without reaching any resolution.

The traditional metaphysician's response is more robust: "This is my idealist system S, which provides a coherent picture of the world, and is consistent with science. It can explain x, y, and z fully. (Moreover, z has always been regarded as extremely difficult – if not impossible.) S can also give partial and provisional answers to p, q, and r." They then challenge the objector: "Let me see your metaphysical system T, which contains your objection as a postulate. We can then compare S and T as a whole in order to evaluate their strengths and weaknesses." The traditional metaphysician thus evaluates and compares systems S and T **as a whole**, to see which is the most coherent and has the greatest explanatory power. In this bake-off the results can be summarised as in Table 2.1. Any individual postulate remains questionable but, so long as it remains part of the strongest system, this is moot.

Metaphors

In this book I have so far avoided metaphors. This is for two reasons. First, the mind-body problem is in a category of its own: unlike anything else. Second, analogies are vague and can be misleading. Now that pan-idealism has been explained, it is the appropriate moment to give a couple of metaphors, and explain pan-idealism in terms of them.

"Inside" versus "outside"

There is a metaphor which says that "from the outside" humans are physical bodies, or, more specifically, human brains; whereas "from the inside" humans are minds. Perhaps this also applies to some higher animals. The panpsychist extension of this is that all individual entities "from the outside" have the character of physical systems; whereas, "from the inside" they are mind-like. This applies even to the fundamental entities of physics.

There are a couple of problems with this "two-sided" picture. First, the analogy is vague. It is unclear whether the "outside view" is fundamental, in which case, this is a type of physicalism. Or the "inside view" might be fundamental, in which case this is a kind of idealism. Or they might be co-equal, in which case this theory is a kind of double-aspect monism. Second, there is an explanatory gap because no explanation has been given as to how these vastly contrasting viewpoints are related to one another.

How does pan-idealism fit in to this picture? Pan-idealism is a specific kind of idealism, so the "inside view" is fundamental. It is also a type of (non-physicalist) panpsychism, because (with trivial qualifications) it agrees with the best current scientific account of the catalogue of the individual entities that exist. In brief – despite its idealism – it is a realist theory about all of the individual entities that exist in the world.

In pan-idealism, the "outside view" does not exist as anything distinct from, or over and above, anything described in the previous paragraph. Instead, the "outside view" is **identically** the maximally correlated structural information combined from the percepts (the "inside views") of **all** experients. "Objective physics" is a synonym for the "outside view", thus defined in entirely mental terms.

"First person" versus "Third person"

This is more or less the above analogy, but phrased differently. Now we say, "from the first person viewpoint", instead of "from the inside". Likewise, we say, "from the third person viewpoint" instead of "from the outside". The "third person viewpoint" is also called the "objective viewpoint" (of physics). **Initially**, the argument proceeds in the same manner:

"From the third person viewpoint" humans are physical bodies, or, more specifically, human brains; whereas "from the first person viewpoint" humans are minds. Perhaps this also applies to some higher animals. The panpsychist extension of this is that all individual entities "from the third person viewpoint" have the character of physical systems; whereas, "from the first person viewpoint" they are mind-like. This applies even to the fundamental entities of physics.

Because this is just a rewording of the previous metaphor, the problems remain the same.

How does pan-idealism fit into this "grammatical persons" picture? Pan-idealism is a specific kind of idealism, so the "first person viewpoint" is fundamental. It is also a type of (non-physicalist) panpsychism, because (with trivial qualifications) it agrees with the best current scientific account of the catalogue of the individual entities that exist. In brief – despite its idealism – it is a realist theory about all of the individual entities that exist in the world.

With this new metaphor, we can characterise objective physics in a novel, and neater way. Pan-idealism does not involve a mythical "impersonal third person viewpoint" at all. Instead, it uses the "second person viewpoint", which is a standard term for "the intersubjective viewpoint". In pan-idealism, the "second person viewpoint" does not exist as anything distinct from, or over and above, anything described in the previous paragraph. Instead, it is **identically** the maximally correlated structural information combined from the percepts (the "first person viewpoints") of **all** experients. "Objective physics" is a synonym for the "second person viewpoint" thus defined.

(Thomas Nagel (1986) has evaluated the third person viewpoint, in both positive and negative terms. His criticisms are relevant to my argument here.)

A toy pan-idealist universe

The purpose here is to give a model world so simple that its associated intersubjective consilience is explicitly determined. It is incomplete because it is static: We will assume that the experients do move in some manner, but (as with a chair in a room in our universe) this motion does not affect the overall situation.

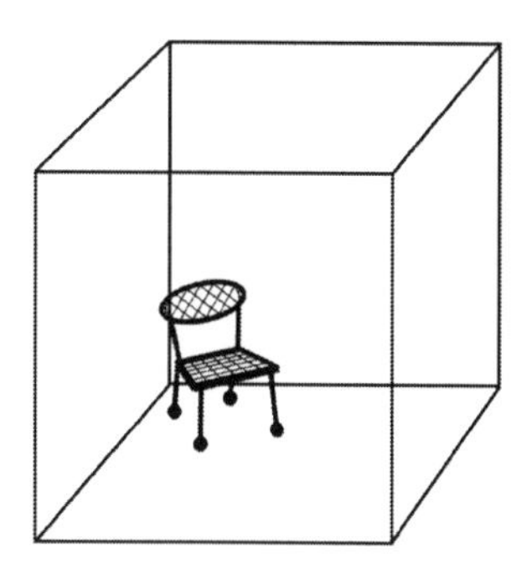

Figure 11.1: The objective physics of a toy pan-idealist universe

Figure 11.1 shows the objective physics of a toy pan-idealist universe, which is in the form of a cube. It contains experients of two types, which I'll call *Light* and *Dark*.

- *Light experients are abundant in the chair's surroundings.* Light experients are invisible to one another, but they can perceive the Dark experients. Their percepts are akin to human visual percepts.
- *Dark experients form the chair.* We are not concerned with the percepts of Dark experients in this example. (Perhaps other Dark experients are invisible to them; and they perceive each Light experient as a speck.)

Figure 11.2 shows the percepts of five, representative Light experients. There are trillions of them and together they perceive the chair from every conceivable distance and angle: (1) sees the chair from directly above; (2) has a close-up view; (3) sees the chair from a different orientation; (4) sees just the leg; and (5) is looking at an empty corner.

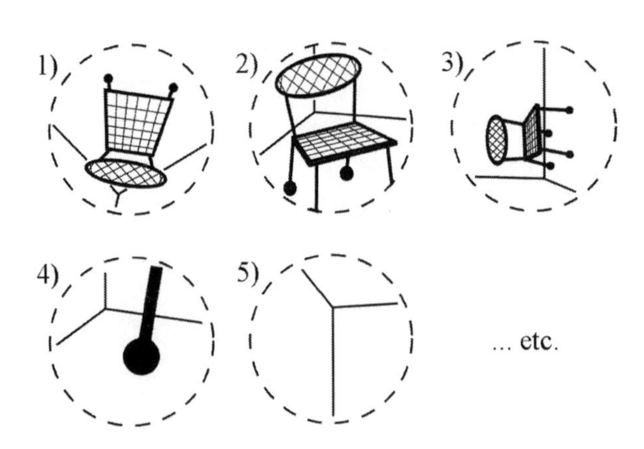

Figure 11.2: The percepts of five, representative Light experients. The percepts of all experients, taken together, are the ontological basis of this pan-idealist universe.

The percepts of all Light and Dark experients constitute the

fundamental ontology in this pan-idealist universe, so Figure 11.2 gives a small sample of this ontology. The objective physics sketched in Figure 11.1 can be **derived** from this basis by maximal intersubjective consilience: that is to say, by combining the percepts of **all** the experients in the most consistent manner possible. In some respects, this is like assembling a single coherent picture from a jigsaw of innumerable pieces. But remember that maximal intersubjective consilience is a mathematical/logical **fact** rather than a **process**.

The objective physics of Figure 11.1 **is nothing other than a compact mathematical summary**, which brings together (combining overall experients) structural information, each piece of which concretely exists within the percept of some particular experient. Objective physics is not something 'extra'. The 3-D shape of the chair is derived in this way. Both the cubical shape of the universe, and the position of the chair within it, can be derived in the same manner: This is because the chair is perceived from all possible viewpoints (within the cube).

The edges of the cube should not really appear in the percepts of Figure 11.2; they are drawn just for our convenience. The above derivation of the universe's shape is made without mentioning edges. If you are concerned that the percepts of many experients, such as (5), are empty, it is possible to modify the toy universe such that the chair is invariably seen. To do this, let the perceptual field of each experient be extended to be an entire sphere, enabling it to perceive other entities in all directions.

Intersubjective consilience in our universe

Pan-idealism holds that the objective physics of our (or any) world is **identically** the maximal intersubjective consilience between all experients. Chapter 8 discussed this at length. Because it is entirely empirical, science-as-a-human-endeavour is entirely based on intersubjective consilience between humans, who are taken to be a representative sample of all experients. There is a strong case that dogs, as higher experients, have percepts passably close to our own; and that this is sufficient to allow them to contribute to the intersubjective consilience.

Chapter 4 showed that quantum systems have the capability to react with different propensities according to the entirety of their environment. This suggests – at least for one who is already a convinced panpsychist on other grounds – that they have significant perceptual abilities. The suggestion is made explicit in a pan-idealist assumption: **Pg** *Experients behave lawfully, with a degree of freedom, according to their percepts.*

Admittedly, in contrast to the toy example, in our world the specific mathematics of intersubjective consilience is almost certainly unknowable. This is because the percepts (in particular their structural information) of all experients are as yet unknown. Even in the case of humans we have to rely on vague verbal reports: it is impossible for you to describe completely and accurately what you are perceiving at this very moment. Despite this unknowability, it is clear that – provided the percepts of all experients are fairly rich – then intersubjective consilience exists, with a definite mathematical form. Moreover, it is appropriate to be identified as the objective physics of the world.

The true and definite (but unknown) mathematical laws of intersubjective consilience that exist within pan-idealism correspond to what are, for the physicalist, the laws of objective physics.

Chalmers' critique of maximal intersubjective consilience

In "Idealism and the Mind-Body Problem" (Seager, 2020, chapter 28), David Chalmers gives, among other things, what is in effect an extended critique of the possibility of maximal intersubjective consilience (pp. 361-62). First, here is Chalmers' characterisation of idealism:

> I will understand idealism broadly, as the thesis that the universe is fundamentally mental, or perhaps that **all concrete facts are grounded in mental facts**.
> (Emphasis added, p. 353)

His argument (pp. 361-62) entails that – because the maximal intersubjective consilience of all experients' percepts of (say) the length of a bone, is not itself experienced by any one of them – "idealism is weakened" (p. 362). He goes on to assert that views similar to maximal intersubjective consilience "are not idealist in the strict sense defined at the start of the chapter, but perhaps are close enough to be interesting" (p. 362).

As discussed in the previous section, maximal intersubjective consilience consists in **specific mathematical rules that combine together – across all experients – structural facts that concretely exist within the percepts of individual experients**. As such, **it is a perfectly concrete fact about the universe**; and it unequivocally satisfies Chalmers' characterisation of idealism: Compare the emphasised words here with the emphasised portion of his definition. This is true even though the precise mathematics of the maximal intersubjective consilience (i.e., the objective physics of the universe) is unknowable.

Is physicalism inevitable?

Some philosophers believe that the successes of science make physicalism well-nigh inevitable. Recall an earlier quotation from Daniel Stoljar:

> Those who deny physicalism are not making a conceptual mistake, but they are, nevertheless, flying in the face not merely of science but also of scientifically informed common sense.
> (Stoljar, 2010, p. 13)

This is also William Seager's view in the following quote, in which, whenever he refers to panpsychism, he means **physicalist** panpsychism:

> There are non-physicalist alternatives to panpsychism. Idealism, for example, retains defenders. ... Arguing for panpsychism over such non-physicalist views cannot be given here, but roughly speaking it is panpsychism's ability to integrate with the scientific view of the world... that [is one of its] main advantages.
> (Brackets and ellipses added, Seager, 2020, p. 9, note 10)

Moreover, Seager is a philosopher who is willing to engage fully with science. See, for example, his chapter, "Consciousness, Information and Panpsychism," in Shear (Ed., 1997), which contains specific mathematical details on adapting quantum theory to physicalist panpsychism.

But pan-idealism is also necessarily consistent with science. One can reach this conclusion 'on the cheap' simply by noting that science is an entirely empirical human activity, wholly based on human percepts and agency. Pan-idealists can thus immediately accept the truth of the entirety of science – including the seemingly utterly materialist-in-tenor neuro- and cognitive sciences.

In this book, however, I have gone further, and have sought to tie pan-idealism to an explicit, contemporary physical theory, GRW, and to demonstrate that they are mutually consistent. Admittedly, GRW has very much an outside chance of being the objective physics of our world. But the fact, which has been demonstrated here, that physicalist GRW and pan-idealist GRW are empirically identical, is more than enough to show that there can be no easy, knock-down proof that pan-idealism is inconsistent with science.

Computers and consciousness

Can computers think? Alan Turing wrote an influential paper in which he argued that this was too vague to be meaningful. Instead, he proposed an objective test, later known as the *Turing Test* (1950). In this, judges hold conversations, on any topics they desire, of say fifteen minutes' duration. These exchanges take place via teletype with two participants, one of which is a human, and the other a suitably programmed computer. If, after many such trials, the judges couldn't reliably tell which participant was the human (say they were correct only 55% of the time), then we must concede that the computer is thinking. (Even if the judges could not distinguish between the computer and human participants, they would still guess correctly 50% of the time.)

It is true that any computer able to pass the Turing Test must be supremely adroit in the scope and flexibility of its logical powers. But does this mean that such a computer is conscious; or is it merely simulating consciousness? Turing was not clear about this. He contended that "instead of arguing continually over this point it is usual to have the polite convention that everyone thinks." And later, "Most of those who support the argument from consciousness could be persuaded to abandon it... They will then probably be willing to accept our test." He admitted that consciousness is not fully understood, but believed that this does not affect the validity of his test. He never explicitly states, however, whether "thinking" – as he himself understands it – necessarily involves consciousness (1950, Section 6(4)).

Today, over seventy years after Turing's paper, no computer has come close to passing his Test. Computers can hold basic conversations on a specific topic. During such tests the judges are instructed not to wander away from this prearranged topic. Even so, the computers frequently make absurd statements that immediately reveal their lack of understanding. Roger Penrose has discussed the contrasting strengths and weaknesses of humans versus computers in performing various tasks (1995, pp. 44-48 and throughout).

Pan-idealism and human consciousness

David Chalmers' famous *Hard Problem* of consciousness – which inspired fresh interest in the mind-body problem – is that of explaining how does it happen that we have conscious, subjective, qualitative experiences (1995, 1996a, 1997).

Pan-idealism affirms that experience is basic, and it is the relevant physics that needs to be explained. Highly developed human consciousness has a holistic character, and this is physically manifest as a quantum entanglement in the brain, which must be sizeable and intricate because our percepts are rich in information. Humans are also agents, possessing libertarian free will. This is also manifest in the entanglement, which describes (rather than constrains) an individual's free propensities to act in a particular way in a given situation. Although our freedom is – far more often than we might like to imagine – limited or even non-existent; on at least some occasions we must possess authentic agency.

Pan-idealism could be proven false, if it is demonstrated that quantum entanglement is irrelevant to the functioning of the human brain in controlling our bodily movements, including speech acts. (It has been argued, by those who deny the possibility of such quantum effects, that the brain is too large, warm, and wet to sustain any such entanglements. To date, these arguments have been brief, and far too sketchy to be decisive. Henry Stapp, working from within an undogmatic Copenhagen perspective, has a well-developed theory of quantum mind (2004a, 2004b, 2007).)

Pan-idealism and digital computers

Now, however, we do have a theory of consciousness. Suppose pan-idealism is true, what does this tell us about the consciousness of a digital computer – even one that passes the Turing Test?

According to pan-idealism, at the most basic level a computer is composed entirely of experients. But these experients play no role in changing the output of the computer. A computer consists of an extremely fast oscillator (the 'clock'), and a vast number of switches. These switches are grouped into memory locations (of size 32, say). There is a special memory location, called the *accumulator*, where all arithmetic operations are carried out. A segment of a program is given in Figure 11.3. The result of this fragment is to decrement the value stored in location 1234 by one. This is followed by a conditional branch.

A digital computer is a deterministic device (despite the conditional jumps). It is entirely classical in the way it works: low-level quantum entanglements within the computer do not influence which instructions are executed; or the computer's output. Its operation is not holistic in any way. This is because its causality is at the level of the instructions, and each instruction depends solely on the contents of the accumulator and of one or two memory locations. Although computers are 'universal' devices, this means only that any digital computer can simulate any other. It does not mean that they can be anything beyond simulators.

Instruction	**Meaning**
LDA 1234	Copy the contents of memory location 1234 into the accumulator
DECA	Decrement (i.e. subtract one from) the contents of the accumulator
STA 1234	Copy the contents of the accumulator into location 1234
JMP0 5678	If the accumulator contains 0, then *jump*: the next instruction to be executed is in location 5678. Otherwise just continue to the next instruction in sequence, *xxx*
xxx	Some arbitrary instruction

Figure 11.3: Fragment of low-level programming language (assembler code) on a digital computer

According to pan-idealists, a digital computer – even one that passed the Turing Test perfectly – is therefore merely simulating consciousness. (If this is correct, then, contrary to Nick Bostrom's contention (2003), we cannot possibly be living in a computer simulation.) However, there is no reason to expect that there could ever be such a successful simulation. David Hodgson, although not a pan-idealist, holds a related position "in which experience/subject/choice is fundamental." His paper "The easy problems ain't so easy" argues that we are wrong to assume that so-called easy problems (such as merely simulating a conversation) "will not be fully solved in advance of the solution to [Chalmers'] hard problem" (Hodgson, 1996, p. 75).

Pan-idealism and quantum computers

Quantum computers are becoming more widespread, and it is now possible to learn about and program them online. You may run your programs, either on a simulator, or (via the Internet) on a real quantum computer. (See https://qiskit.org/ for instruction; and also https://quantum-computing.ibm.com/, where you can set up a free account.)

Whereas computers use bits, which are either 0 or 1, quantum computers use qubits (quantum bits) which can take any value between 0 and 1. More precisely, *qubits* have the physical characteristics of spin-½ particles (though they are usually implemented differently). Defining $|0\rangle = |z\uparrow\rangle$ and $|1\rangle = |z\downarrow\rangle$ a general qubit $|q\rangle$ is given by

$$|q\rangle = c|0\rangle + d|1\rangle \qquad (1)$$

Here c and d are complex, with $|c|^2 + |d|^2 = 1$. Recall the discussion of spinors towards the end of chapter 4. Quantum computers use the $(|0\rangle, |1\rangle)$-basis (i.e., the z-basis) by default, but it is possible to work in any basis.

To-date, quantum computers possess a limited number of qubits (a few dozen is a ballpark figure). The power of entanglement makes up for this lack of numbers. No matter how they are physically implemented, qubits must be kept at a temperature near absolute zero, to prevent entanglements being lost. The bulk of a quantum computer is therefore a refrigeration device, and the qubits are contained within a few chips hidden deep inside.

Perhaps the easiest way to program a quantum computer is via a graphical interface, as shown on the left of Figure 11.4. It is analogous to musical notation. For each qubit to be used in a program there is a horizontal line, akin to a staff line in a musical score, and with time flowing from left to right. There are also (the same number of) classical lines associated with each qubit. The classical lines are always depicted together, as symbolised by the double line at the bottom of the Figure.

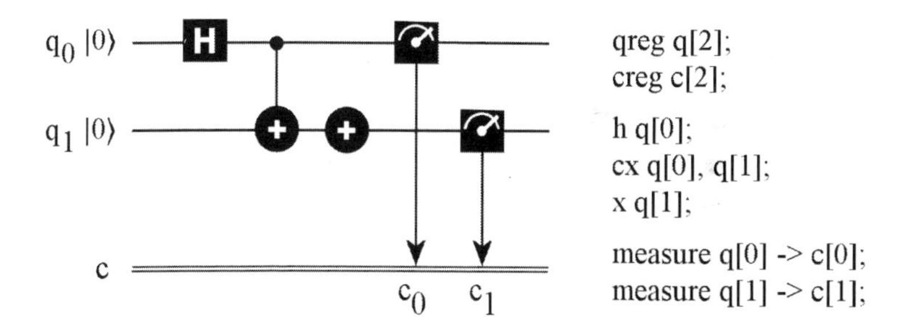

Figure 11.4: Program for a quantum computer. Bell entanglement.

The first task is to choose the number of qubits you wish to use. Many of the computers at the IBM site have five qubits, labelled as q_0 through q_4, drawn from top to bottom in the Figure. In the example, we are considering a Bell-type entanglement, and for this purpose only two qubits, $|q_0\rangle$ $|q_1\rangle$, and two corresponding classical lines, c_0 c_1, are needed. At the start of each program run, every qubit is automatically initialised to $|0\rangle$.

The various icons in the program signify operations carried out on qubits. In creating the program, these icons are dragged from a palette and dropped into place. When the program is run, the operations will be carried out in sequence from left to right. They will not be explained in full here: for complete details, see the manuals and teaching materials at the Qiskit and IBM websites mentioned above.

The first icon is the *Hadamard gate* [H]. Its effect on a qubit that is in state $|0\rangle$, here $|q_0\rangle$, is to turn it into the (z-basis) superposition

$$|q_0\rangle = \frac{1}{\sqrt{2}}(|0\rangle + |1\rangle) \qquad (2)$$

(If you look at the discussion of spinors in chapter 4, you will see that this is $|x\uparrow\rangle$, which is one of the basis spinors in the x-basis (note well: this is x not z). This example shows that the concept of being in a superposition is not absolute, but is relative to a given basis. From now on, however, whenever I use the unqualified term *superposition,* I mean superposition in the favoured z-basis.)

The second icon is the *Controlled NOT gate* (also known as the CNOT or CX gate). This joins a pair of qubits. The small (•) end of the icon is the *control qubit,* and the other, larger ('hot cross bun') end is what I will call the *dependent qubit*. In the simple situation where the control qubit is either $|0\rangle$ or $|1\rangle$, CNOT has the following effect:

> *If the control qubit is* $|1\rangle$ *then apply* NOT *to the dependent qubit; otherwise do nothing.* (The NOT gate will be explained in a moment.)

But in our example the control qubit $|q_0\rangle$ is in a superposition, as in (2) above; moreover, the dependent qubit $|q_1\rangle = |0\rangle$.

Although it will not be explained here, the effect of the CNOT gate in this situation is to entangle $|q_0\rangle$ and $|q_1\rangle$ in such a way that, if z-measurements were to be carried out on both qubits immediately afterwards, then invariably either both $|q_0\rangle$ and $|q_1\rangle$ would be $|0\rangle$, or both would be $|1\rangle$. The online texts explain that this entanglement can be written formally as

$$|q_1q_0\rangle = \frac{1}{\sqrt{2}}|00\rangle + 0\,|01\rangle + 0\,|10\rangle + \frac{1}{\sqrt{2}}|11\rangle \qquad (3)$$

$$= \frac{1}{\sqrt{2}}(|00\rangle + |11\rangle)$$

The third icon ('hot cross bun') is a *NOT gate* (or X gate). In the simplest cases, its effect is to convert $|0\rangle$ to $|1\rangle$, and $|1\rangle$ to $|0\rangle$. In general, it converts the superposition $c|0\rangle + d|1\rangle$ to $d|0\rangle + c|1\rangle$. I have included this final gate to make the entanglement more akin to those found in the Aspect-type experiments described in chapter 5.

For this particular example, when the qubits are measured after applying this gate, if one is found to be $|0\rangle$, then the other will be $|1\rangle$; but we cannot predict which qubit will have which value. The online texts explain that the final entanglement can be written formally as

$$|q_1q_0\rangle = 0\,|00\rangle + \frac{1}{\sqrt{2}}|01\rangle + \frac{1}{\sqrt{2}}|10\rangle + 0\,|11\rangle \qquad (4)$$

$$= \frac{1}{\sqrt{2}}(|01\rangle + |10\rangle)$$

All of the above operators are reversible. Moreover, they are each their own inverse, so performing any one of them twice in succession has the same effect as doing nothing.

The measurement icons: Measuring a qubit produces one classical bit of information (either 0 or 1). This is recorded on the classical wire that corresponds to the qubit. With few exceptions, measurement is not reversible because it collapses the wavefunction. One exception is that the immediate repetition of a measurement on the same qubit gives the same result, and leaves the entire quantum state of the system unaltered (as it was since the first measurement).

In Figure 11.4, $|q_0\rangle$ is measured first, and the result is recorded on c_0. As a result of this, the quantum state of equation (4) will have collapsed (with equal probability) to

$$\text{EITHER } |q_1q_0\rangle = |01\rangle \qquad \text{OR } |q_1q_0\rangle = |10\rangle \qquad (5)$$

In the EITHER case, c_0 will record the $|q_0\rangle$ value as 1; in the OR case, c_0 will record the $|q_0\rangle$ value as 0. These statements are true because $|q_0\rangle$ is associated with the rightmost position within the ket: $|\bullet q_0\rangle$.

Finally, $|q_1\rangle$ is measured, and the result is recorded on c_1. This measurement cannot collapse either of the wavefunctions given in (5) further. Can you explain why the result of this second measurement (recorded on c_1) is always the opposite of the first? (Hint: consider the EITHER and OR cases separately.)

Computer code: The fragment of programming code to the right of Figure 11.4 was automatically generated by the graphical interface. Understanding it is not essential because you can run the program without it. But it is useful in teaching you about the programming language when a graphical interface is unavailable. It is also possible to generate the graphic to the left of the Figure by using commands within the programming code.

The code is straightforward once you have learned its rules:

1. Declare the number of qubits and corresponding classical bits you will be using
2. State in chronological order which gates are to be used,

and which qubits they operate on

For the CNOT gate (called the cx gate in the language) it is important to specify the qubits in the order you intend. The qubit written first is the control qubit. If the qubits had been given in the reverse order, then the CNOT icon would appear inverted in Figure 11.4. This latter is a perfectly legitimate, but distinct, operation; it is not what is required for our example.

3. Make the measurements

Each measurement is sent to the corresponding classical bit. Occasionally it is appropriate to make a measurement before the end of the program; but beware that doing this typically destroys information.

Running the program: Once your program (either graphical or coded) is complete it can be run on a simulator, or on a real quantum computer, or both. In each run, the program is executed many times: there might be a total of say 1,000 trials. After all this, statistics are gathered summarising the results. In our example, the result of each trial is a two-bit binary number.

On a simulator the overall outcome might be that 521 trials resulted in 01, and 479 resulted in 10. These are automatically plotted as a histogram. The numbers are not exactly equal because each trial is, in effect, a coin toss. There are many additional graphical tools for presenting the results. Simulators model perfect machines that follow quantum rules exactly.

On a real quantum computer, there might be a dozen or so trials resulting in 00 or 11. This is due to imperfections in the machine (say caused by thermal noise). Another disadvantage is that, because the real computers are in high demand, there are sometimes longer delays in getting results back.

Quantum computers and pan-idealism: Are the quanta within a quantum computer conscious? According to pan-idealism they are – at least in the trivial sense that they are centres of experience. But it is important not to overstate this: they have zero cognitive powers (and so they are not conscious in Penrose's sense). Moreover, as these machines are currently designed, they cannot perceive, nor do they have agency over, anything external to the computer. Because these quantum systems are so simple, and because the initial state is fully known, the computer program, acting on the qubits, can control them fully. There is thus a slight analogy between the way in which this is done and with Frankfurt's sinister Mr Green (see **Objection 5** in chapter 9). Two points of disanalogy here are, first that the human brain is highly complex, and second that its complete quantum state is never fully known at any given time. The first point might be overcome, but only in the far distant future. It is unlikely that the second difficulty will ever be surmounted.

There is a philosophical thought experiment called *the brain in a vat*. This imagines that you are in reality a brain trapped in a vat, with a scientist feeding you false information about your environment by artificial signals sent through your sensory nerves. The philosophical question (first raised by Plato in his famous parable of "The Cave"; and later, in different terms, by Descartes) is: Could this – admittedly utterly implausible – situation possibly be true? I will pass over this question for the case of humans. In the case of quantum entanglements within present-day quantum computers, these are slightly analogous to minds (better described as mere experients) in a vat. Again, it is impossible to overstress how extremely primitive they are. In line with the vat metaphor, their perceptual states at every moment are fully controlled by the program, moreover they have no ability to interact with the environment of the computer.

How many experients are present during runs of a quantum computer? Pan-idealism gives an answer: At the beginning of each computer trial, each qubit is individually set to $|0\rangle$, and so each is a separate, primitive experient. As qubits become entangled, by this very fact, they coalesce into one experient. Likewise, as they disentangle, they separate into distinct experients.

It is too soon to speculate how far quantum computing technology will develop in the distant future. There might be quantum computers built on entirely different principles from those that exist today.

Farewell

Thank you for reading this book. I hope you **don't** conclude that "it fills a much-needed gap in the literature (!)"; but instead you find it to be a novel addition to the tiny (though hopefully growing) section on idealism within your vast library on the mind-body problem. I remain convinced of the truth of pan-idealism, despite its prima facie weirdness.

Bibliography

If a URL is given in addition to the reference, then page numbers given in the text refer to the document as it appears on the Internet.

Aspect, A., Dalibard, J. & Roger, G. (1982). "Experimental test of Bell's inequalities using time-varying analyzers". *Physical Review Letters*, **49** (25), pp. 1804-07

Audi, R. (Ed.) (1999, 2nd edition). *The Cambridge Dictionary of Philosophy* (Cambridge: Cambridge University Press)

Ayer, A. (1936/1971). *Language, Truth and Logic* (Harmondsworth: Pelican Books)

Ayer, A. (1972). *Russell* (London: Fontana)

Baggott, J. (1992). *The Meaning of Quantum Theory* (Oxford: Oxford University Press)

Baldwin, T. (2004). *Stanford Encyclopedia of Philosophy: Moore* http://plato.stanford.edu/entries/moore#6

Becker, A. (2018). *What is Real? The Unfinished Quest for the Meaning of Quantum Mechanics* (London: John Murray)

Bell, J. (1964). "On the Einstein Podolsky Rosen Paradox". *Physics*, **1** (3), pp. 195-200

Bell, J. (1966). "On the Problem of Hidden Variables in Quantum Mechanics". *Reviews of Modern Physics*, **38** (3), pp. 447-52

Bell, J. (1981a). "Bertlemann's socks and the nature of reality". *Journal de Physique, Colloque C2, Suppl. 3*, **42** (3), pp. 41-62

Bell, J. (1981b). "Quantum mechanics for cosmologists", Ch. 27 in Isham, C., Penrose, R. & Sciama, D. (Eds.), *Quantum Gravity 2* (Oxford: Oxford University Press), pp. 611-37

Bell, J. (1982). "On the impossible pilot wave". *Foundations of Physics*, **12** (10), pp. 989-99

Bell, J. (1987). "Are there quantum jumps?", in *Schrödinger: Centenary of a Polymath* (Cambridge: Cambridge University Press), pp. 1-14

Bell, J. (1989). "Six Possible Worlds of Quantum Mechanics", in *Proceedings of the Nobel Symposium 65: Possible Worlds in Humanities, Arts and Sciences* (held in Stockholm, 11-15 August 1986), Sture, A. (Ed.) (Walter de Gruyter), pp. 359-73

Bell, J. (1990). "Against 'Measurement'". *Physics World*, (August) pp. 33-40

Bell, J. (2004, 2nd edition). *Speakable and Unspeakable in Quantum Mechanics* (Cambridge: Cambridge University Press) [A collection of Bell's key papers, introduced by Alain Aspect.]

Bell, M., Gottfried, K. & Veltman, M. (Eds.) (2001). *John S. Bell on the Foundations of Quantum Mechanics* (London: World Scientific) [A very similar collection to the above, edited by Bell's widow Mary, and others.]

Beller, M. (1999). *Quantum Dialogue: The Making of a Revolution* (Chicago: University of Chicago Press)

Beller, M. & Fine, A. (1994). "Bohr's Response to EPR", in Faye, J. & Folse, H. (Eds.)

Berkeley, G. (1710 & 1713 respectively/1996). *Principles of Human Knowledge* and *Three Dialogues* (Oxford: Oxford University Press)

BIPM [Bureau International des Poids et Mesures] (2019, 9th edition). *SI Brochure: The International System of Units (SI)*. (Downloaded 19 January 2021): https://www.bipm.org/en/publications/si-brochure/SI Brochure-9-concise-EN.pdf

Block, N., Flanagan, O. & Güzeldere, G. (Eds.) (1997). *The Nature of Consciousness: Philosophical Debates* (Cambridge, MA: MIT Press)

Bohm, D. (1951/1989). *Quantum Theory* (New York: Dover)

Bohm, D. (1980). *Wholeness and the Implicate Order* (London: Routledge)

Bohm, D. & Hiley, B. (1993). *The Undivided Universe: an ontological interpretation of quantum theory* (London: Routledge)

Bohr, N. (1935). "Can Quantum-Mechanical Description of

Physical Reality be considered Complete?" *Physical Review*, **48** (October 15), pp. 696-702

Bolender, J. (2001). "An argument for idealism". *Journal of Consciousness Studies*, **8** (4), pp. 37-61

Born, M. (1971). *The Born-Einstein Letters* (London: Macmillan)

Bostrom, N. (2003). "Are you living in a computer simulation?" *Philosophical Quarterly*, **53** (211), pp. 243-55

Bouwmeester, D., Ekert, A. & Zeilinger, A. (Eds.) (2000). *The Physics of Quantum Information* (Berlin: Springer)

Brüntrup, G. & Jaskolla, L. (Eds.) (2017). *Panpsychism: Contemporary Perspectives* (Oxford: Oxford University Press)

Chalmers, D. (1995). "Facing up to the problem of consciousness". *Journal of Consciousness Studies*, **2** (3), pp. 200-219

Chalmers, D. (1996a). *The Conscious Mind: In Search of a Fundamental Theory* (Oxford: Oxford University Press)

Chalmers, D. (1996b). "David Chalmers on Neural Correlates of Consciousness (Tucson 1996)" (Video of conference talk, accessed 15 May 2020) https://www.youtube.com/watch?v=uMFkuUPP_XY

Chalmers, D. (1997). "Moving forward on the problem of consciousness". *Journal of Consciousness Studies*, **4** (1), pp. 3-46

Churchland, P. (1996). "The Hornswoggle Problem". *Journal of Consciousness Studies*, **3** (5-6), pp. 402-408

Clarke, D. (2003). *Panpsychism and the Religious Attitude* (New York: SUNY Press)

Clarke, D. (2004). *Panpsychism: Past and Recent Selected Readings* (New York: SUNY Press)

Clauser, J., Horne, M., Shimony, A. & Holt, R. (1969). "Proposed Experiment to Test Local Hidden-Variable Theories". *Physical Review Letters*, **23** (15), pp. 880-84

Clauser, J. & Shimony, A. (1978). "Bell's theorem. Experimental tests and implications". *Reports on Progress in Physics*, **41**, pp. 1881-1927

Clayton, P. (2004). *Mind and Emergence: From Quantum to Consciousness* (Oxford: Oxford University Press)

Cottingham, J. *et al.* (Eds.) (1991). *The Philosophical Writings of Descartes: Volume III, The Correspondence* (Cambridge: Cambridge University Press)

Cottingham, J. (Ed.) (2008, 2nd edition). *Western Philosophy: An Anthology* (Oxford: Blackwell)

Crick, F. (1994). *The Astonishing Hypothesis: The Scientific Search for the Soul* (London: Simon & Schuster)

Crick, F. & Koch, C. (1990). "Towards a neurobiological theory of consciousness". *Seminars in the neurosciences*, **2**, pp. 263-275

Crick, F. & Koch, C. (1992). "The problem of consciousness". *Scientific American*, September 1992, pp. 110-117

Damour, T. & Burniat, M. (2017). *Mysteries of the Quantum Universe* (London: Penguin Books)

Davies, P. & Brown, J. (Eds.) (1986). *The Ghost in the Atom* (Cambridge: Cambridge University Press)

de Quincey, C. (2002). *Radical Nature: Rediscovering the Soul of Matter* (Vermont: Invisible Cities Press)

Dennett, D. (1991). *Consciousness Explained* (Boston: Little, Brown)

Dennett, D. (1992). "The unimagined preposterousness of zombies". *Journal of Consciousness Studies*, **2** (4), pp. 322-33

Dickson, M. (2001). "The EPR Experiment: A Prelude to Bohr's Reply to EPR" (arXiv:quant-ph/0102053v1 9 February 2001). Downloaded from: https://arxiv.org/abs/quant-ph/0102053v1

Dunham, J., Grant, I. & Watson, S. (2011). *Idealism: The History of a Philosophy* (Durham: Acumen)

Dürr, D., Goldstein, S. & Zanghì, N. (2013). *Quantum Physics Without Quantum Philosophy* (Berlin: Springer)

Eddington, A. (1928). *The Nature of the Physical World* (Cambridge: Cambridge University Press)

Einstein, A. (1920/1970). *Relativity: The Special and the General Theory* (London: Methuen & Co.) [Original German edition, 1916]

Einstein, A., Podolsky, B. & Rosen, N. (1935). "Can Quantum-Mechanical Description of Physical Reality be Considered Complete?" *Physical Review*, **47** (May 15), pp. 777-80

Ells, P. (2011). *Panpsychism: The Philosophy of the Sensuous Cosmos* (Winchester: John Hunt Publishing)

Ells, P. (2018). "Alternatives to Physicalism", in Castro, J., Fowler, B. & Gomes, L. (Eds.), *Time, Science and the Critique of Technological Reason: Essays in Honour of Hermínio Martins* (Switzerland: Palgrave Macmillan)

Everett, H. (1957). "'Relative State' Formulation of Quantum Mechanics". *Review of Modern Physics,* **29**, pp. 454-462

Faye, J. & Folse, H. (Eds.) (1994). *Niels Bohr and Contemporary Philosophy* (Dordrecht: Kluwer Academic Publishers)

Feynman, R. (1990). QED: *The Strange Theory of Light and Matter* (London: Penguin)

Fine, A. (1986). *The Shaky Game: Einstein, Realism and the Quantum Theory* (Chicago: University of Chicago Press)

Foster, J. (1982). *The Case for Idealism* (London: Routledge & Kegan Paul)

French, S. (2014). *The Structure of the World: Metaphysics and Representation* (Oxford: Oxford University Press)

Ghirardi, G. (2007, revised edition). *Sneaking a Look at God's Cards: Unraveling the Mysteries of Quantum Mechanics* (Princeton: Princeton University Press)

Ghirardi, G., Rimini, A. & Weber, T. (1986). "Unified dynamics for microscopic and macroscopic systems". *Physical Review D*, **34** (2) (July 15), pp. 470-91

Goff, P. & Papineau, D. (1967). "Consciousness & Physicalism" (YouTube debate: published 16 November 2017, accessed 29 April 2020)

Goldschmidt, T. & Pearce, K. (Eds.) (2017). *Idealism: New Essays in Metaphysics* (Oxford: Oxford University Press)

Goodman, N. (1978). *Ways of Worldmaking* (Indianapolis: Hackett)

Gregory, R. (Ed.) (1987). *The Oxford Companion to the Mind* (Oxford: Oxford University Press)

Gregory, R. (1998, 5th edition). *Eye and Brain: The Psychology of Seeing* (Oxford: Oxford University Press)

Griffin, D. (1998). *Unsnarling the World-Knot: Consciousness, Freedom and the Mind-Body Problem* (Berkeley, CA: University of California Press)

Griffiths, D. & Schroeter, D. (2018, 3rd edition). *Introduction to Quantum Mechanics* (Cambridge: Cambridge University Press)

Hawking, S. (1988). *A Brief History of Time: From the Big Bang to Black Holes* (London: Bantam)

Hawking, S. (2011). "Why Are We Here? | Stephen Hawking | Google Zeitgeist" (Talk: published 18 May 2011; accessed 17 May 2020) https://www.youtube.com/watch?v=r4TO1iLZmcw

Hawking, S. & Ellis, G. (1973). *The large scale structure of space-time* (Cambridge: Cambridge University Press)

Heisenberg, W. (1949/1930). *The Physical Principles of the Quantum Theory* (New York: Dover)

Heisenberg, W. (1958/2000). *Physics and Philosophy* (London: Penguin Classics)

Hodgson, D. (1991). *The Mind Matters: Consciousness and Choice in a Quantum World* (Oxford: Oxford University Press)

Hodgson, D. (1996). "The easy problems ain't so easy". *Journal of Consciousness Studies*, **3** (1), pp. 69-75

Horgan, J. (1999). *The Undiscovered Mind: How the Brain Defies Explanation* (London: Phoenix)

Hume, D. (1748/1999). *An Enquiry concerning Human Understanding* (Oxford: Oxford University Press)

Hut, P., Alford, M. & Tegmark, M. (2006). "On Math, Matter and Mind". *Foundations of Physics*, **36**, pp. 765-794 http://arxiv.

org/pdf/physics/0510188

James, W. (1890/1983). *The Principles of Psychology* (Cambridge, MA: Harvard University Press)

James, W. (1909/2005). *A Pluralistic Universe* (Stilwell, KS: Digireads)

Kane, R. (1998). *The Significance of Free Will* (Oxford: Oxford University Press)

Kane, R. (Ed.) (2002a). *Free Will* (Oxford: Blackwell)

Kane, R. (Ed.) (2002b). *The Oxford Handbook of Free Will* (Oxford: Blackwell)

Kant, I. (A1781/B1787/1998). *Critique of Pure Reason* (Cambridge: Cambridge University Press)

Kim, J. (1996, 1st edition). *Philosophy of Mind* (Oxford: Westview Press)

Koch, C. (2004). "Thinking about the conscious mind" (Review of Searle, 2004). *Science*, **306**, pp. 979-80

Leibniz, G. (1714). *Monadology*. English translation (downloaded 29 April 2020) taken from: http://www.marxists.org/reference/subject/philosophy/works/ge/leibniz.htm

Levine, J. (1983). "Materialism and qualia: the explanatory gap". *Pacific Philosophical Quarterly*, **64**, pp. 354-61

Lewis, P. (2016). *Quantum Ontology: A Guide to the Metaphysics of Quantum Mechanics* (Oxford: Oxford University Press)

Lorimer, D. (Ed.) (2004). *Science, Consciousness & Ultimate Reality* (Exeter: Imprint Academic)

Lycan, W. (Ed.) (1999, 2nd edition). *Mind and Cognition: An Anthology* (Oxford: Blackwell)

Macleod, C. (2016). *Stanford Encyclopedia of Philosophy: John Stuart Mill*, downloaded 6 September 2020: https://plato.stanford.edu/entries/mill/

Mattuck, R. (1992). *A Guide to Feynman Diagrams in the Many-Body Problem* (New York: Dover)

Maudlin, T. (2012). *Philosophy of Physics: Space and Time* (Princeton: Princeton University Press)

Maudlin, T. (2014). "What Bell Did". *J. Phys. A: Math. Theor.*, **47** (October 8), pp. 1-24

Maudlin, T. (2019). *Philosophy of Physics: Quantum Theory* (Princeton: Princeton University Press)

McGinn, C. (1999). *The Mysterious Flame* (Oxford: Basic Books)

McLaughlin, B. & Bartlett, G. (2004). "Have Noë and Thompson cast doubts on the neural correlates of consciousness programme?" *Journal of Consciousness Studies*, **11** (1), pp. 56-67

Mermin, D. (1990). *Boojums All the Way Through: Communicating Science in a Prosaic Age* (Cambridge: Cambridge University Press)

Mermin, D. (1993). "Hidden variables and the two theorems of John Bell". *Reviews of Modern Physics*, **65** (3), pp. 803-15

Misner, C., Thorne, K. & Wheeler, J. (1973). *Gravitation* (San Francisco: Freeman)

Nagel, T. (1979). *Mortal Questions* (Cambridge: Cambridge University Press)

Nagel, T. (1986). *The View From Nowhere* (Oxford: Oxford University Press)

Noë, A. & Thompson, E. (2004). "Are There Neural Correlates of Consciousness?" *Journal of Consciousness Studies*, **11** (1), pp. 3-28

Norsen, T. (2017). *Foundations of Quantum Mechanics: An Exploration of the Physical Meaning of Quantum Theory* (Springer Nature)

Pais, A. (1979). "Einstein and the quantum theory". *Reviews of Modern Physics*, **51** (4), pp. 863-914

Pais, A. (2005, 2nd edition). *Subtle is the Lord: The science and the life of Albert Einstein* (Oxford: Oxford University Press)

Papineau, D. & Selina, H. (2000). *Introducing Consciousness* (Cambridge: Icon Books)

Parthey, C., Matveev, A., Alnis, J. *et al.* (2011). "Improved Measurement of the Hydrogen 1S-2S Transition Frequency".

Phys. Rev. Lett., **107** (20), 203001, Epub

Peat, D. (1997). *Infinite Potential: The Life and Times of David Bohm* (New York: Basic Books)

Penrose, R. (1989). *The Emperor's New Mind: Concerning Computers, Minds and the Laws of Physics* (Oxford: Oxford University Press)

Penrose, R. (1995). *Shadows of the Mind: A Search for the Missing Science of Consciousness* (London: Vintage)

Penrose, R. (2004). *The Road to Reality: A Complete Guide to the Laws of the Universe* (London: Jonathan Cape)

Poland, J. (1994). *Physicalism: The Philosophical Foundations* (Oxford: Clarendon Press)

Rauch, D., Handsteiner, J., Hochrainer, A. *et al.* (2018). "Cosmic Bell Test Using Random Measurement Settings from High-Redshift Quasars". *Physical Review Letters*, **121** (August 20, 080403), pp. 1-4

Robinson, D. (2008). *Consciousness and Mental Life* (New York: Columbia University Press)

Romanes, G. (1895). *Mind and Motion and Monism* (London: Longmans, Green & Co.) Downloaded from: www.gutenberg.org

Rosenberg, G. (2004). *A Place for Consciousness: Probing the Deep Structure of the Natural World* (Oxford: Oxford University Press)

Russell, B. (1927/2007). *The Analysis of Matter* (Nottingham: Russell Press)

Savile, A. (2005). *Kant's Critique of Pure Reason: An Orientation to the Central Theme* (Oxford: Blackwell)

Schilpp, P. (Ed.) (1949/2000, 3rd edition). *Albert Einstein: Philosopher-Scientist* (Peru, IL: Open Court)

Schopenhauer, A. (Trans. Payne, E.) (1859/1966). *The World as Will and Representation: Volumes I & II* (New York: Dover)

Schrödinger, E. (1935a). "Discussion of Probability Relations between Separated Systems". *Proc. Camb. Phil. Soc.* **31** (10), pp. 555-63

Schrödinger, E. (1935b). "The Present Situation in Quantum Mechanics" [See Trimmer, J. (1980)]

Schrödinger, E. (1936). "Probability Relations between Separated Systems". *Proc. Camb. Phil. Soc.* **32** (10), pp. 446-52

Schrödinger, E. (1949). "Erwin Schrödinger – 'Do Electrons Think?' (BBC 1949)". (Accessed 24 July 2020) https://www.youtube.com/watch?v=hCwR1ztUXtU

Schrödinger, E. (2014/1948/1951). *Nature and the Greeks* and *Science and Humanism* (Cambridge: Cambridge University Press)

Seager, W. (1995). "Consciousness, information and panpsychism". *Journal of Consciousness Studies,* **2** (3), pp. 272-88

Seager, W. (1999). *Theories of Consciousness: An Introduction and Assessment* (London: Routledge)

Seager, W. (2005). *Stanford Encyclopedia of Philosophy: Panpsychism* http://plato.stanford.edu/entries/panpsychism

Seager, W. (Ed.) (2020). *The Routledge Handbook of Panpsychism* (London: Routledge)

Searle, J. (1984). *Minds, Brains and Science* (Cambridge, MA: Harvard University Press)

Searle, J. (1992). *The Rediscovery of the Mind* (Cambridge, MA: MIT Press)

Searle, J. (1997). *The Mystery of Consciousness* (London: Granta Books)

Searle, J. (2002). "Why I am not a property dualist". *Journal of Consciousness Studies,* **9** (12), pp. 57-64

Searle, J. (2004). *Mind: A Brief Introduction* (Oxford: Oxford University Press)

Shear, J. (Ed.) (1997). *Explaining Consciousness – The 'Hard Problem'* (Cambridge, MA: MIT Press)

Skrbina, D. (2003). "Panpsychism as an Underlying Theme in Western Philosophy – A Survey Paper". *Journal of*

Consciousness Studies, **10** (3), pp. 4-46

Skrbina, D. (2005). *Panpsychism in the West* (Cambridge, MA: MIT Press)

Skrbina, D. (Ed.) (2009). *Mind that Abides* (Philadelphia, PA: John Benjamins)

Soon, S., Brass, M., Heinze, H-J & Haynes, J-D (2008). "Unconscious determinants of free decisions in the human brain". *Nature Neuroscience*, **11** (5), pp. 543-45

Sprigge, T. (1983). *The Vindication of Absolute Idealism* (Edinburgh: Edinburgh University Press)

Stapp, H. (1993, 1st edition). *Mind, Matter and Quantum Mechanics* (Berlin: Springer)

Stapp, H. (1996). "The Hard Problem: A Quantum Approach". *Journal of Consciousness Studies*, **3** (3), pp. 194-210

Stapp, H. (2004a, 2nd edition). *Mind, Matter and Quantum Mechanics* (Berlin: Springer)

Stapp, H. (2004b). "Quantum Leaps in Philosophy of Mind: Reply to Bourget's Critique". *Journal of Consciousness Studies*, **11** (12), pp. 43-49

Stapp, H. (2007). *Mindful Universe: Quantum Mechanics and the Participating Observer* (Berlin: Springer)

Stoljar, D. (2015 March 9 revision [original 2001]). *Stanford Encyclopedia of Philosophy: Physicalism*, downloaded 29 April 2020: http://plato.stanford.edu/entries/physicalism

Stoljar, D. (2010). *Physicalism* (London: Routledge)

Stone, M. & Wolff, J. (Eds.) (2000). *The Proper Ambition of Science* (London: Routledge)

Strawson, G. (1994). *Mental Reality* (Cambridge, MA: MIT Press)

Strawson, G. (2006a). "Realistic Monism: Why Physicalism Entails Panpsychism". *Journal of Consciousness Studies*, **13** (10-11), pp. 3-31

Strawson, G. (2006b). "Panpsychism?" *Journal of Consciousness Studies*, **13** (10-11), pp. 184-280

Strawson, G. (2006c). *Consciousness and its place in nature* (Exeter:

Imprint Academic)

Strawson, G. (2009a). "Realistic Monism: Why Physicalism Entails Panpsychism", in Skrbina, D. (Ed.), *Mind that Abides*, pp. 33-56 (Philadelphia, PA: John Benjamins)

Strawson, G. (2009b). "Appendix to Realistic Monism: Why Physicalism Entails Panpsychism", in Skrbina, D. (Ed.), *Mind that Abides*, pp. 57-63 (Philadelphia, PA: John Benjamins)

Susskind, L. & Friedman, A. (2014). *Quantum Mechanics: The Theoretical Minimum* (Penguin)

Tegmark, M. (2003). "Parallel universes" in Barrow, J., Davies, P. & Harper, C. (Eds.), *Science and Ultimate Reality: Quantum Theory, Cosmology, and Complexity*, honouring John Wheeler's 90th birthday (Cambridge: Cambridge University Press) http://space.mit.edu/home/tegmark/multiverse.pdf

Trimmer, J. (1980). "The Present Situation in Quantum Mechanics: A Translation of Schrödinger's 'Cat Paradox' Paper". *Proc. Amer. Phil. Soc.*, **124** (5), pp. 323-38

Turing, A. (1950). "Computing machinery and intelligence". *Mind*, **49**, pp. 433-460

Vimel, R. (2009). "Meanings attributed to the term 'consciousness'". *Journal of Consciousness Studies*, **16** (5), pp. 9-27

Von Neumann, J. (1932/1971). *Mathematical Foundations of Quantum Mechanics* (New Jersey: Princeton University Press) (This English translation from the original German was approved by the author.)

Walter, S. & Heckmann, H-D (Eds.) (2003). *Physicalism and Mental Causation: The Metaphysics of Mind and Action* (Exeter: Imprint Academic)

Whitaker, A. (1996). *Einstein, Bohr and the Quantum Dilemma* (Cambridge: Cambridge University Press)

Whitaker, A. (2016). *John Stewart Bell and Twentieth-Century Physics: Vision and Integrity* (Oxford: Oxford University

Press)

Whitehead, A. (1929/1978). *Process and Reality: An Essay in Cosmology – Corrected Edition* (New York: Free Press)

ACADEMIC AND SPECIALIST

Iff Books publishes non-fiction. It aims to work with authors and titles that augment our understanding of the human condition, society and civilisation, and the world or universe in which we live.
If you have enjoyed this book, why not tell other readers by posting a review on your preferred book site.
Recent bestsellers from Iff Books are:

Why Materialism Is Baloney
How true skeptics know there is no death and fathom answers to life, the universe, and everything
Bernardo Kastrup
A hard-nosed, logical, and skeptic non-materialist metaphysics, according to which the body is in mind, not mind in the body.
Paperback: 978-1-78279-362-5 ebook: 978-1-78279-361-8

The Fall
Steve Taylor
The Fall discusses human achievement versus the issues of war, patriarchy and social inequality.
Paperback: 978-1-78535-804-3 ebook: 978-1-78535-805-0

Brief Peeks Beyond
Critical essays on metaphysics, neuroscience, free will, skepticism and culture
Bernardo Kastrup
An incisive, original, compelling alternative to current mainstream cultural views and assumptions.
Paperback: 978-1-78535-018-4 ebook: 978-1-78535-019-1

Framespotting

Changing how you look at things changes how you see them

Laurence & Alison Matthews

A punchy, upbeat guide to framespotting. Spot deceptions and hidden assumptions; swap growth for growing up. See and be free.

Paperback: 978-1-78279-689-3 ebook: 978-1-78279-822-4

Is There an Afterlife?

David Fontana

Is there an Afterlife? If so what is it like? How do Western ideas of the afterlife compare with Eastern? David Fontana presents the historical and contemporary evidence for survival of physical death.

Paperback: 978-1-90381-690-5

Nothing Matters

a book about nothing

Ronald Green

Thinking about Nothing opens the world to everything by illuminating new angles to old problems and stimulating new ways of thinking.

Paperback: 978-1-84694-707-0 ebook: 978-1-78099-016-3

Panpsychism

The Philosophy of the Sensuous Cosmos

Peter Ells

Are free will and mind chimeras? This book, anti-materialistic but respecting science, answers: No! Mind is foundational to all existence.

Paperback: 978-1-84694-505-2 ebook: 978-1-78099-018-7

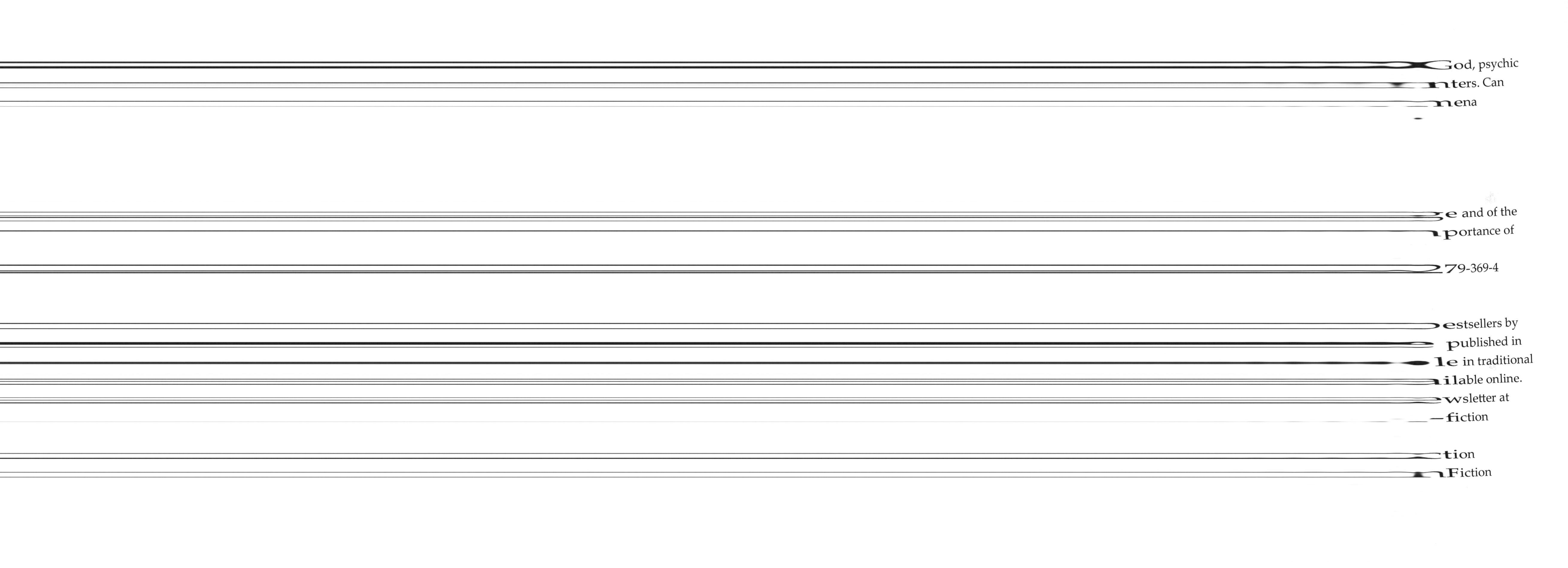

od, psychic
nters. Can
nena

e and of the
portance of

79-369-4

estsellers by
published in
le in traditional
ilable online.
wsletter at
-fiction

tion
Fiction

Punk Science

Inside the Mind of God

Manjir Samanta-Laughton

Many have experienced unexplainable phenomena; God, psychic abilities, extraordinary healing and angelic encounters. Can cutting-edge science actually explain phenomena previously thought of as 'paranormal'?

Paperback: 978-1-90504-793-2

The Vagabond Spirit of Poetry

Edward Clarke

Spend time with the wisest poets of the modern age and of the past, and let Edward Clarke remind you of the importance of poetry in our industrialized world.

Paperback: 978-1-78279-370-0 ebook: 978-1-78279-369-4

Readers of ebooks can buy or view any of these bestsellers by clicking on the live link in the title. Most titles are published in paperback and as an ebook. Paperbacks are available in traditional bookshops. Both print and ebook formats are available online.

Find more titles and sign up to our readers' newsletter at http://www.johnhuntpublishing.com/non-fiction

Follow us on Facebook at https://www.facebook.com/JHPNonFiction

and Twitter at https://twitter.com/JHPNonFiction